Environmental Factors
In Respiratory
Disease

ENVIRONMENTAL SCIENCES

An Interdisciplinary Monograph Series

EDITORS

DOUGLAS H. K. LEE
National Institute of
Environmental Health Sciences
Research Triangle Park
North Carolina

E. WENDELL HEWSON
Department of
Atmospheric Science
Oregon State University
Corvallis, Oregon

DANIEL OKUN
University of North Carolina
Department of Environmental
Sciences and Engineering
Chapel Hill, North Carolina

ARTHUR C. STERN, editor, AIR POLLUTION, Second Edition. In three volumes, 1968

L. FISHBEIN, W. G. FLAMM, and H. L. FALK, CHEMICAL MUTAGENS: Environmental Effects on Biological Systems, 1970

DOUGLAS H. K. LEE and DAVID MINARD, editors, PHYSIOLOGY, ENVIRONMENT, AND MAN, 1970

KARL D. KRYTER, THE EFFECTS OF NOISE ON MAN, 1970

R. E. MUNN, BIOMETEOROLOGICAL METHODS, 1970

M. M. KEY, L. E. KERR, and M. BUNDY, PULMONARY REACTIONS TO COAL DUST: "A Review of U. S. Experience," 1971

DOUGLAS H. K. LEE, editor, METALLIC CONTAMINANTS AND HUMAN HEALTH, 1972

DOUGLAS H. K. LEE, editor, ENVIRONMENTAL FACTORS IN RESPIRATORY DISEASE, 1972

H. ELDON SUTTON and MAUREEN I. HARRIS, editors, MUTAGENIC EFFECTS OF ENVIRONMENTAL CONTAMINANTS, 1972

In preparation

MOHAMED K. YOUSEF, STEVEN M. HORVATH, and ROBERT W. BULLARD, PHYSIOLOGICAL ADAPTATIONS: Desert and Mountain

DOUGLAS H. K. LEE and PAUL KOTIN, editors, MULTIPLE FACTORS IN THE CAUSATION OF ENVIRONMENTALLY INDUCED DISEASE, 1972

Fogarty International Center Proceedings No. 11

Environmental Factors In Respiratory Disease

Scientific Editor
Douglas H. K. Lee

NATIONAL INSTITUTE OF
ENVIRONMENTAL HEALTH SCIENCES
RESEARCH TRIANGLE PARK, NORTH CAROLINA

Sponsored by
National Institute of Environmental Health Sciences
Research Triangle Park, North Carolina
and
John E. Fogarty International Center
National Institutes of Health
Bethesda, Maryland

Academic Press
New York and London 1972

ACADEMIC PRESS, INC.
111 Fifth Avenue, New York, New York 10003

United Kingdom Edition published by
ACADEMIC PRESS, INC. (LONDON) LTD.
24/28 Oval Road, London NW1 7DD

LIBRARY OF CONGRESS CATALOG CARD NUMBER: 75-189162

CONTENTS

CHAPTER 8. SOME SPECIFIC EFFECTS OF ENVIRONMENTAL AGENTS

Benjamin G. Ferris

CHAPTER 9. ENVIRONMENTAL FACTORS IN CHRONIC LUNG DISEASE

John P. Wyatt

CHAPTER 10. HOST VARIABLES IN PULMONARY RESPONSES TO THE ENVIRONMENT

Gareth M. Green

CHAPTER 11. INTERACTION OF INFECTIOUS DISEASE AND AIR POLLUTANTS: INFLUENCE OF "TOLERANCE"

David L. Coffin

CHAPTER 15. ENVIRONMENTAL FACTORS IN BRONCHIAL ASTHMA
Carl M. Shy, Victor Hasselblad, Leo T. Heiderscheit, and Arlan A. Cohen

CHAPTER 16. TOWARDS AN OPTIMUM ENVIRONMENT 237
Eric J. Cassell

CHAPTER 17. CONCLUSIONS AND RESERVATIONS
Douglas H. K. Lee

ORGANIZING PANEL

Peter G. Condliffe, Chief, Conference and Seminar Program Branch, Fogarty International Center, National Institutes of Health, Bethesda, Maryland 20014

Kaye H. Kilburn, Director, Division of Environmental Medicine, Duke University Medical Center, Durham, North Carolina 27706

Douglas H. K. Lee, Associate Director, National Institute of Environmental Health Sciences, Research Triangle Park, North Carolina 27709

Jay A. Nadel, Associate Professor, Department of Medicine, Cardiovascular Research Institute, University of California Medical Center, San Francisco, California 94122

Irving J. Selikoff, Environmental Science Laboratory, Mt. Sinai School of Medicine, New York, New York 10029

John P. Wyatt, Professor and Chairman of Pathology, University of Manitoba School of Medicine, Winnipeg, Canada

CONTRIBUTORS

Eric J. Cassell, Department of Public Health, Cornell University Medical College, New York, New York 10021

David L. Coffin, National Environmental Research Center, Environmental Protection Agency, Research Triangle Park, North Carolina 27711

Benjamin G. Ferris, Department of Physiology, Harvard School of Public Health, Boston, Massachusetts 02115

G. F. Filley, University of Colorado School of Medicine and Webb–Waring Lung Institute, Denver, Colorado 80220

Aron B. Fisher, Departments of Medicine and Physiology, University of Pennsylvania School of Medicine, Philadelphia, Pennsylvania 19104

H. Kenneth Fisher, Department of Medicine, Harborview Medical Center, University of Washington, Seattle, Washington 98104

Robert Frank, Department of Environmental Health, University of Washington School of Public Health and Community Medicine, Seattle, Washington 98105

Gareth M. Green, University of Vermont College of Medicine, Burlington, Vermont 05401

E. Cuyler Hammond, American Cancer Society, New York, New York 10017

Robert J. M. Horton, National Environmental Research Center, Environmental Protection Agency, Research Triangle Park, North Carolina 27711

P. Macklem, Respiratory Division, Royal Victoria Hospital, McGill University, Montreal, Canada

Paul E. Morrow, Department of Radiation Biology and Biophysics, University of Rochester School of Medicine and Dentistry, Rochester, New York 14620

Norton Nelson, Institute of Environmental Medicine, New York University Medical Center, New York, New York 10016

Carl M. Shy, National Environmental Research Center, Environmental Protection Agency, Research Triangle Park, North Carolina 27711

Norman C. Staub, Cardiovascular Research Institute and Department of Physiology, University of California, San Francisco, California 94122

EDITORIAL COMMENT

As governmental and public interest in environmental quality increases, and as conservation or remedial programs are planned, a number of scientists are called upon for advice or decision in areas involving fields beyond their own personal expertise. Program managers, experienced in administration, also find themselves in need of information on technical matters that fall within their jurisdiction.

The announcement that a United Nations Conference on human environment is to be held in Stockholm in June, 1972 pointed out the fact that these needs are worldwide. It stimulated the John E. Fogarty International Center of the National Institutes of Health to investigate how these needs might be met. In conjunction with the National Institute of Environmental Health Sciences, also part of the National Institutes of Health, a decision was made to prepare four books on the aspects of environmental health for which suitable résumés were not readily available.

Environmental factors in respiratory disease was the topic selected for the third of the set. As for the other volumes, a small panel of experts in the field was asked to delimit the scope that should be followed, to indicate the specific topics within that scope, and to suggest other experts who could contribute to the volume. Contributors brought draft papers to a three-day Workshop where the drafts were thoroughly discussed, amendments suggested, and integration between chapters developed. Extensive editing followed.

It has not been easy to preserve a balance between simplification for the non-specialist and adequacy as viewed by the expert. The text now appearing has been checked by the contributors, but I must accept responsibility for any undue selectivity that may have occurred, as well as for errors of omission or commission. I hope, however, that the text will help those who need to know the state of current knowledge on the health significance of environmental factors in respiratory disease but who do not have the time to pursue the detailed literature or to seek a compilation directed to their special needs.

Douglas H. K. Lee

CONCENTRATION UNITS AND CONVERSION FACTORS

Metric and proportional units are used somewhat indiscriminately for indicating the concentration of toxic agents. In this volume the units preferred by the individual contributor have been retained. The reader who wishes to make comparisons between concentrations expressed in different units will find the following data useful.

In solid and liquid mixtures, proportional units refer to weights and are easily converted to metric equivalents:

$$1 \text{ part per million (ppm)} = 1 \text{ mg/kg}$$
$$1 \text{ part per billion (ppb)} = 1 \text{ } \mu\text{g/kg}$$
$$(1 \text{ kg of liquid of density } 1 = 1 \text{ li})$$

In gaseous mixtures, however, proportional units refer to volumes, so that conversion varies with the molecular weight of the dispersed substance, temperature, and barometric pressure. The conversion formula is

$$1 \text{ part per million by volume}$$
$$\text{at } 25°C \text{ and } 760 \text{ mm Hg pressure} = 0.041 \text{ (molecular weight) mg/m}^3$$

Note: The factor 0.041 represents $273/(298 \times 22.4)$

Particulate matter in air is often expressed in terms of millions of particles per cubic foot or per cubic meter. To convert from cubic foot to cubic meter, multiply by 35.3. (This also gives particles per cubic centimeter.)

CHAPTER 1. INTRODUCTION:
A CHANGING ENVIRONMENT —
A PROBLEM FOR ALL

NORTON NELSON, Institute of Environmental Medicine,
New York University Medical Center, New York

Although concern for environmental factors as possible contributors to disease is of long standing, it is only within recent years that the field has received the widespread attention of the medical and scientific community that it has now achieved. A rather detailed study has been directed to occupational respiratory disease in recent decades, and the personal habit, cigarette smoking, has received shorter although considerable interest. Community air pollution has recently played a significant role in bringing increased attention to this field.

In a general sense, two trends have been preeminent in lending increased importance to the study of environmental factors and disease. Included in the first are the rapid development of technology with its associated pollutants from home heating, from automobiles, and from industry. These, in association with an increasing urbanization have made the problem of air pollution dramatically worse and more widespread. The second trend comprises all those advances which have led to substantial improvements in health and survival, such as the control of infectious diseases through antibiotics and immunization, better sanitation and water supplies, and improved nutrition. Progress in the conquest of major killing and maiming diseases has permitted us to give merited attention to some of the newer diseases of the environment.

Thus diseases of environmental origin, such as those believed to result from air pollution, are now receiving a level of concern and an intensity of study which are several orders of magnitude greater than they have received in the past. Many of our first rank medical scientists are engaged in this research and are finding the resolution of etiologic and pathogenic mechanisms highly challenging scientific questions.

In this introductory chapter a brief examination will be made of some trends and developments which bear on our concern with respiratory disease and the environment. As noted above, these can be regarded as falling into two areas: changing patterns in population and health, and changing patterns in technology.

Inasmuch as the field of environmental health is primarily concerned with possible adverse health effects arising out of technology, an awareness of the changing patterns of technology is essential. Indeed, the field of environmental health should be regarded as potentially in a state of constant flux, since technology itself is changing rapidly. The point has frequently been made that cultural evolution (for which one might substitute technical evolution) progresses at an exponential rate. This has been observed in primitive as well as modern technologies. An example of the increasing pace is provided by the observation that, whereas early in this century the interval between the development of a concept and its technological application was estimated at 37 years, it is estimated to average now only 14 years.

When this point of the acceleration of technological evolution is made, it is often coupled with the statement that biological evolution, by contrast, progresses at only an arithmetic rate. The statement that biological evolution proceeds slowly can be taken to imply that man's intrinsic biologic capabilities and weaknesses are not changing rapidly. This does not, however, (and this is important) exclude circumstances which can alter the population composition, such as: a) major migration of ethnic groups; b) the influence of selective mechanisms that once weeded out more sensitive segments of the population through uncontrolled diseases of infancy and childhood that are now no longer so prominent; and c) possibly new mutational patterns.

CHANGING PATTERNS IN HEALTH

The population at risk today is a very different one from that of a century ago or even a quarter of a century ago. In 1860 the median age of the population was 19.4, in 1960 it was 29.6. Life expectancy at birth in 1900 was 50 years, in 1960, 70 years (1). To a considerable extent, of course, this represents a lowering in deaths resulting from the diseases of infancy and early youth. Improvements in life expectancy after the age of 30 have been fewer. These circumstances lead to a different age distribution. Thus there is a greater percentage of the population now surviving into the older and probably more susceptible age groups than in former times, and also more people surviving to develop late effects of slowly acting agents such as carcinogens.

Alterations in disease patterns have been extensive, perhaps especially in those involving the respiratory system. Thus, whereas in 1930, before the availability of antibiotics, the annual death rate from lobar pneumonia was 45.3 per 100,000, by 1960 it had dropped to 5.6 per 100,000. Similarly, while tuberculosis in 1900 was responsible for 194 deaths per 100,000, by 1950 the rate had declined to 22.5, and in 1967 to 3.5 per 100,000 (2). On the other hand, while chronic bronchitis and emphysema were held responsible for a combined death rate of 10 per million in 1950 in New York, by 1965 the rate had risen to 70 per million -- a seven-fold increase in only fifteen years (3). Receiving major attention has been the incidence of lung cancer, which has also skyrocketed in recent decades, increasing about seven-fold over the thirty-year period 1930-1960 (4).

Environmental factors have come under strong suspicion as playing important roles in several of these diseases, particularly chronic bronchitis, emphysema, and lung cancer. Cigarette smoking and community air pollution have both been considered as possible contributors. These issues will be examined in detail later in this volume.

CHANGING PATTERNS IN TECHNOLOGY

The environmental exposure of the population is related to the social patterns and to the technology which supports the population. These circumstances will emerge more fully as we examine some of the changing patterns in technology and other factors that alter environmental exposures.

Changing patterns of environmental factors will be discussed under four headings: 1) the occupational environment, 2) the community environment, 3) personal habits, and 4) the home.

Occupational Exposures

Aside from the direct interest that it has engendered in occupational disease itself, experience in this field has provided a major source of instruction for environmental health in general. Major and acknowledged improvements in the cleanliness of the industrial environment have come in the last several decades, primarily through the institution of better control procedures. Nevertheless, it is still often true that industrial groups have higher exposures than do the general consumer or other special groups. This generally higher exposure of workers persists in spite of major improvements in containment of hazardous materials. Containment is rarely complete and, even where it is normally successful, breakdowns occur which frequently provide an opportunity for exposure of maintenance and repair staff to toxic agents. Thus, the occupational setting retains its importance as a source of general information in the study of environmental disease.

In broad terms, the major patterns in industry appear to be increasing automation, with the consequence that the number of workers per unit of production tends to decrease. Some chemical plants now operate with a small fraction of the number of employees per ton of production required several decades ago. Paralleling this trend is the increase in those employed in service industries. Service personnel form an increasingly significant part of our work force and one that has received very little attention in respect to special occupational hazards. An opposite trend is evident in the proportion of the working force engaged in agriculture, which continues to decline. Agriculture is increasingly losing its distinction from heavy industry as agricultural production tends to be undertaken in larger and larger units, and under management principles that are very similar to those of major corporations. As one specific example, pesticide application is more and more in the hands of the specially trained and (hopefully) qualified persons, and this trend will become even more pronounced.

The "dusty trades" were among the earliest to respond to the demand for a cleaner work environment. In spite of this, however, the response has often been sluggish and incomplete. Silicosis and asbestosis are still occurring, and although we were aware thirty years ago of the lung cancer hazards from asbestos and radon daughters, the needed clean-up has been disgracefully laggard. A brand new dust hazard was presented recently in the addition of enzymes to detergents. Since pre-testing requirements are woefully scanty, it is possible that other new dust hazards will go undetected until adverse effects appear with continued exposure.

Although many of the old hazardous chemicals continue to be produced, e.g., chromium chemicals, benzene, and benzidine, others are being phased out, either because their importance has declined or because they are too difficult to live with (e.g., beta napthylamine).

The rapid growth of chemical manufacture is illustrated in Table 1. The enormous proliferation of agricultural chemicals is shown also by the circumstances that there are now some 60,000 registrations for pesticide formulations on file in the Department of Agriculture. The proliferation of food additives is similarly indicated by the 5,000 registrations with the Food and Drug Administration.

New chemicals used as intermediates in the manufacture of other chemical products also are emerging. Two important examples are the isocyanates used in the manufacture of polyurethane, associated with an interesting respiratory syndrome, and the halo ethers used as intermediates in the production of resins and other products. The latter have recently been reported to be very potent lung

TABLE 1 -- PRODUCTION OF SYNTHETIC ORGANIC CHEMICALS (5)

(Millions lbs. per year)

	1938	1958	1966
Plastics	130	4,500	13,585
Synthetic Rubber	5	2,200	3,929
Surface Active Agents	-	1,355	3,321
Insecticides and Agricultural Chemicals*	8	540	1,013

*not including fertilizers

carcinogens (6). The production of both has increased substantially in recent years.

We have entered a new period in the U. S. with the new Federal legislation -- Occupational Safety and Health Act of 1970. The new laws should bring about better and more systematic control and fuller access to the study of occupational exposures. This in turn should lead to the development of a better understanding of the implications of these occupational exposures. Such knowledge will also be useful in understanding lesser exposures of the general public to the same agents.

Community Air Pollution

Pollutants of importance to the residents of the community are occasionally of a highly specialized nature, representing a unique air contaminant with a restricted neighborhood distribution; e.g., beryllium. Of wider interest are the more universally distributed agents arising from such ubiquitous sources as automobiles, power plants, and space heating.

It is essential to maintain alertness to the possibility that specialized or exotic industrial effluents may present neighborhood health hazards. However, major attention needs to be directed to the changing patterns of general community air pollution. Pollution of the old "Pittsburgh" or "St. Louis

6

type", characterized by dense clouds of black smoke and heavy soot and fly ash contamination, is virtually gone in most parts of this country. This is the result of improved combustion techniques, better fuels, and the installation of dust-collecting equipment in discharge stacks.

A later trend, now fully evident in a number of places, is a decline in sulfur oxide concentration resulting from restrictions on the sulfur content of fuels. It can be expected that restrictions on sulfur content of fuel will become more widespread and more stringent. However, the limited amount of low sulfur fuels limits the extent to which this can be effected. Although there is a moderate amount of research aimed at the removal of sulfur from fuel before use, and from stack gases after combustion, these developments have not proceeded far enough to have significant impact now or for the next few years.

Despite improvement in some areas in respect to sulfur oxides, the increasing demand for power, which continues to grow exponentially (Table 2), suggests that sulfur oxide concentrations, now declining in some areas, may reach a plateau or even start an upward trend in other areas unless additional steps are taken. It appears that we will be dependent on fossil fuels longer than had been anticipated. This results from a lag in availability of nuclear power, due in part to a slower production of equipment than had been projected, and in part to strong public objections. In respect to the latter, clearly what is needed is a mechanism for better determination of total power needs and a public understanding of the trade-offs between alternative sources.

TABLE 2 -- PROJECTION OF SOURCE OF ELECTRIC POWER (megawatts) (5)

Source Year	Hydro-electric	Fossil	Nuclear
1960	33,000	150,000	1,000
1980	70,000	220,000	150,000
2000	90,000	850,000	750,000

Automobiles continue to increase more rapidly than the population (Figure 1). It is the automobile which has been chiefly responsible for the "Los Angeles" or oxidizing type of smog, which depends primarily on the nitrogen dioxide and hydrocarbon emissions from automobiles. In addition, the automobile is usually the chief source of carbon monoxide. Although the output of pollutants per unit of power is being reduced, the increasing number of automobiles, and increasing horsepower, are, in many areas, keeping the total emission at previous levels or even continuing its increase.

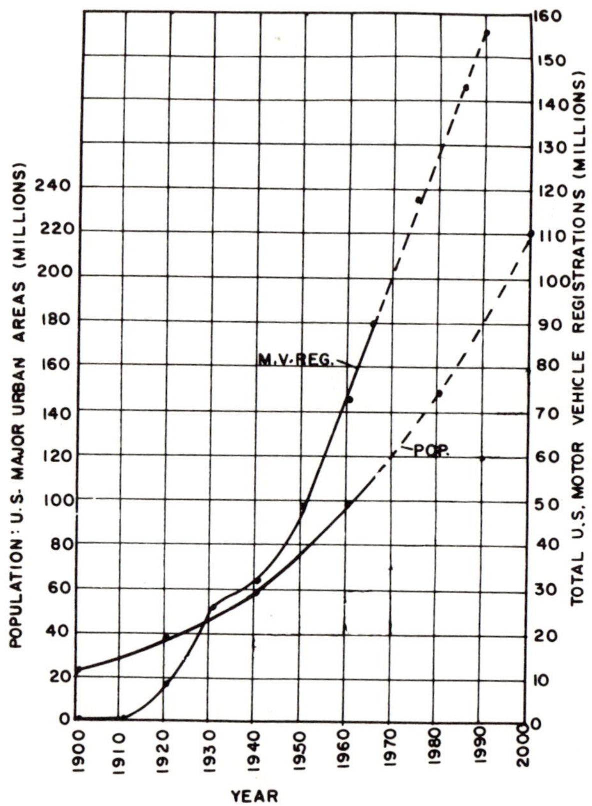

Fig. 1. U. S. Urban Population Motor Vehicle Registrations, 1900-2000 (5).

Personal Habits

The question of personal habits and their relationship to respiratory disease will be briefly mentioned. There is an encouraging trend in the number of adults who are giving up cigarette smoking. Similarly encouraging is a modest indication that, in certain segments of society, cigarette smoking is tending to start among the young at a later age. One should not be highly optimistic about these modest trends, for cigarette consumption remains extremely high. It is thus quite obvious that the pathology resulting from cigarette smoking will be around for many years to come. Clearly what is required is to make cigarette smoking entirely unfashionable among the young.

Home Environment

The health hazards present in the home environment have not in the past received the attention they deserve. With the proliferation of new household chemicals under the kitchen sink and aerosol sprays of every description, the problem becomes even more acute. The possible relationship of enzyme detergents to emphysema or respiratory sensitization should certainly be considered. Similarly, the growing tendency to apply everything from rug cleaners to hair dyes as aerosols should make us all inquire what the consequences of inhaling this witch's brew may be.

COMMENT

The lung merits the special attention of those concerned with noxious factors in the environment. It presents the largest surface which the body exposes to the external environment, a surface far larger in its total area than the skin. The respiratory system has well-developed protective mechanisms, particularly in the conducting airways through which the air reaches the deep lungs; however, the ultimate surfaces of exposure, the alveolar membranes, are extremely delicate and fragile and are highly vulnerable to toxic agents which reach them. The conducting airways, though more substantial, are by no means insensitive to environmental trauma and this system has its own special pathogenic response patterns.

The consequences of exposure to external hazards range from trivial to lethal -- with nuisance, annoyance, and reversible impairment at one end of the scale; and disablement, chronic disease, and death at the other. An objective assessment of the extent of adverse health effects from environmental factors is a difficult and complicated process; this assessment generally requires the full interplay of laboratory and epidemiological approaches.

It is of the greatest urgency that reliable estimates of the role of environmental factors in respiratory disease be secured as a basis for rational control action. In some instances the suspect pollutant can be controlled at a trivial cost, easily acceptable to society; in these cases, obviously, control should be instituted and the problem solved. However, this is generally not the case. In most instances a point is reached in seeking improvements where the cost becomes extremely burdensome. Ignorance, moreover, can lead to costly and unnecessary control of an innocuous agent, while the noxious one is overlooked. A wise choice between alternative technologies is decisively important for reasons not merely of economy but also of health. It is only through the reliable assessment of health hazards from environmental contaminants that priorities can be soundly selected.

The rest of this volume will examine in depth issues briefly touched on here, as well as others not mentioned. Part I (Chapters 2-5) deals with the basic structure and function of the lung and its defensive mechanisms. Part II (Chapters 6-11) discusses the responses that the lung structure makes to various environmental insults and the ultimate effects upon its efficiency. Part III (Chapters 12-15) takes up some special problems that have been causing concern. A final chapter (Chapter 16) summarizes the current situation on environmental factors in respiratory disease and emphasizes the need for research where our present knowledge and understanding are inadequate.

Finally, it should be pointed out that much of the work presented in the following relates to studies currently under way and still to be completed. Many of the conclusions set forth are, therefore, preliminary and will require further work for their confirmation.

REFERENCES

1. Goerke, L. S. and Stebbins, E. L. (1968). Mustard's Introduction to Public Health. MacMillan, New York.

2. U. S. Bureau of the Census. (1969). Statistical Abstract of the United States: 1969. (90th ed.) Washington, D. C.

3. Densen, P. M. et al. (1967). A survey of respiratory disease among New York City postal and transit workers. Env. Res. 1: 265.

4. U. S. National Center for Health Statistics. (1968). Vital Statistics Rates in the United States, 1940-1960. Public Health Service Pub. 1677. Supt. of Documents, U. S. Govt. Print. Off.

5. Nelson, N. et al. (1970). Environmental health. In: P. Handler, (ed.). Biology and the Future of Man. Oxford University Press, New York.

6. Laskin, S. et al. (1971). Tumors of the respiratory tract induced by inhalation of bis(chloromethyl)ether. Arch. Env. Health. 23: 135.

PART I

THE LUNG AND ITS BASIC REACTIONS

CHAPTER 2. ANATOMY AND PHYSIOLOGY OF THE LUNG

NORMAN C. STAUB, Cardiovascular Research Institute and
Department of Physiology, University of California,
San Francisco, California

This paper is a survey of pulmonary anatomy and physiology as it relates to the problem of environmental factors in respiratory disease. For more information there are numerous books and reviews on lung structure (1-4) and on physiology (5-8).

The main function of the lung is to enable air to come into close proximity to the red blood cells flowing in the pulmonary capillaries, so that oxygen (O_2) may be taken into the blood and carbon dioxide (CO_2) given off. The degree of success in accomplishing the main function can be gauged by measuring the arterial blood O_2 and CO_2 pressures, P_aO_2 and P_aCO_2, respectively.

The structural adaptations needed to produce the required functional relationship include development of the right ventricle and pulmonary vascular system, by which all blood returning from the body tissues to the heart is pumped through a vast array of thin sheets in the lung, topologically infolded into a relatively small space (9). The available internal surface area of the adult human lung is approximately 1 m^2/kg body weight at 75% of total lung capacity (TLC). The large internal surface area is not related to the act of distributing air in the lung (ventilation, $\dot{V}$) but to the need to distribute a large quantity of blood (perfusion, $\dot{Q}$) in a very thin film. This is done in such a manner that even under stressful circumstances: a) the capillary transit time, during which gas exchange occurs, is long enough to permit equilibration; b) the thickness of tissue through which gas diffuses is minimized; c) the resistance to flow in the pulmonary vascular bed remains low.

AIRWAYS

The gross and microscopic structure of the airways is well known and there are good quantitative data on the branching pattern, cross sectional area, volume, and other derived parameters for the human lung (7). Intensive physiologic study in recent years has vastly improved our understanding of the airway function, although there is still a great deal to learn.

The airways can be classified longitudinally into three groups: the cartilaginous airways (trachea and bronchi), the membranous airways (bronchioles), and the gas exchange airways (respiratory bronchioles and alveolar ducts).

CARTILAGINOUS AIRWAYS

Structure

Cartilaginous airways include the trachea and all its branchings down to the small bronchi (approximately 0.1 - 0.3 cm in diameter at TLC). The main structural supports for the bronchi are incomplete cartilaginous rings and plates connected by a strong fibrous layer. Within this layer is a layer of circularly arranged smooth muscle whose motor innervation is via the parasympathetic branch of the autonomic nervous system (vagus nerve). The sympathetic branch of the autonomic nervous system is inhibitory, but whether it directly innervates the airway smooth muscle is not known. The smooth muscle is also capable of being contracted (narrowing the airway) by direct stimulation. The innermost layer is the mucous membrane.

The surface of the mucous membrane is pseudostratified, ciliated epithelium. Interspersed between the ciliated cells are mucus-secreting goblet cells. A distinctive feature of the large cartilagenous bronchi is the presence of bronchial glands situated deep within the mucous membrane connecting to the surface through ducts. These glands contain mucous and serous secreting elements and may be stimulated via the vagus nerve. The bronchial glands are particularly dense in the medium size bronchi, decreasing in numbers and size distally, and finally disappearing altogether near the transition from cartilaginous bronchi to membranous bronchioles.

The surface cells are separated from the air in the bronchi by a thin muco-ciliary blanket, formed by continuous serous and mucous secretion of the airway epithelium and glands. The blanket continually moves upward through the airways to the pharynx propelled by the constantly beating cilia (1,000 strokes per minute). Little is known about the mechanisms regulating the activity of the cilia. External innervation is not necessary.

The sensory innervation of the large airways includes a variety of undifferentiated fibers whose receptor endings appear to be adjacent to the surface epithelial layer. These fibers, most of which run to the central nervous system through the vagus nerve, are sensitive to irritation and appear to be identical to the cough receptors. These receptors are particularly abundant in the trachea and largest bronchi.

The blood supply of the bronchi is via the bronchial circulation which arises from the aorta (systemic circulation).

Function

The cartilaginous airways are the conduits through which gas enters and leaves the respiratory gas exchange areas of the lung. Since these airways do not allow gas exchange across their walls, they contribute to that portion of the freshly inspired breath that does not undergo gas exchange; the dead space. Indeed, the cartilaginous airways, together with the nose and throat, account for almost all of the anatomical dead space volume. In addition, because of the air flow through this system of branching and narrowing tubes, the cartilaginous airways contribute a large portion of the total resistance to breathing. There is a reciprocal relationship between the dead space volume and the airway resistance (10). The resistance to gas flow in the bronchi can be increased by reflex or local stimulation of the surface smooth muscle which causes constriction (narrowing) of these airways. Both inert particles impinging on the airway surface and chemically active agents may stimulate the irritant constrictor mechanism.

The cartilaginous airways also aid in the warming and humidification of incoming air, particularly during mouth breathing when the nasal mechanism is by-passed. In addition, foreign inhaled particles in the size range between 2 and 10 microns usually settle or impact onto the muco-ciliary surface layer during inspiration, and are transported out of the airways to the pharynx where they are swallowed. Highly soluble gaseous materials (e.g., sulfur dioxide) are very efficiently scrubbed out of the inspired air; none reaches the delicate gas exchange surfaces directly.

Chronic irritant stimuli appear to cause hypertrophy of the mucous secreting elements and glands of the upper airway with a parallel increase in the production and secretion of airway mucus (11,12).

MEMBRANOUS AIRWAYS

Structure

The bronchioles are direct continuations of the cartilaginous airways. They end in the terminal bronchioles (approximately 0.06 cm in diameter). Bronchioles differ from bronchi chiefly in the absence of the fibrocartilage framework and of secretory glands.

The structural support of the bronchioles comes from their being directly imbedded into the connective tissue framework of the lung. They contain a circular continuous layer of smooth muscle whose innervation is believed similar to that of bronchial muscle

The mucosa is continuous with that of the bronchi. The epithelial layer consists of low, ciliated, columnar cells which gradually become more cuboidal as the terminal bronchiole is approached.

Interspersed with the ciliated cells is a decreasing number of goblet cells, which normally disappear completely at the terminal bronchiole.

The blood supply to the bronchioles is via the bronchial artery (systemic circulation). The sensory innervation consists mainly of undifferentiated fibers in the subepithelial portion of the mucosa.

Figure 1 shows a current lung model with the proper anatomical relationships among the various elements. The model may aid in understanding more of the relations of structure and function in the lung (3,13).

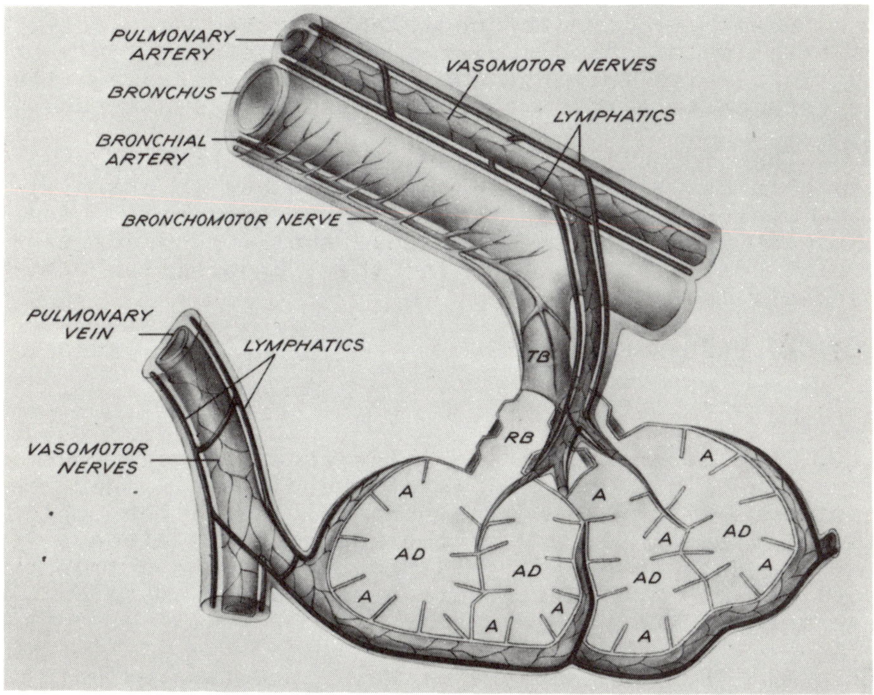

Fig. 1. Lung model incorporating currently known structure-function relationships. Three classes of airways as discussed in text are shown. Two terminal respiratory units are depicted. Note the relations of the bronchial and pulmonary circulations, lymphatics, and motor innervation. RB = respiratory bronchiole; TB = terminal bronchiole; AD = alveolar ducts; A = alveoli.

Function

The difference between small bronchi and large bronchioles is not as sharp as the anatomical description suggests. Actually, there is a gradual transition. The bronchioles contribute to the dead space and air flow resistance. Normally, the contribution is considerably less than that of the bronchi (14). Both the dead space and the portion of airway resistance located in the bronchioles are affected by lung volume, since these airways are imbedded in the lung tissue and are passively dilated during lung inflation and narrowed in lung deflation (expiration).

Edema of the bronchiolar mucosa, or active contraction of the circular smooth muscle reflexly or by local irritation, can greatly increase the resistance to airflow through these very small tubes.

The mucociliary blanket in the terminal bronchioles is mainly a watery fluid but it gradually accumulates mucus as it moves up the airways. Under chronic irritant stimulation, the mucus-secreting goblet cell population in the bronchioles may markedly increase.

GAS EXCHANGE AIRWAYS

Structure

The respiratory bronchioles and alveolar ducts are the final branchings of the airways. They are continous with the terminal bronchioles. The chief distinguishing feature of the gas exchange airways is the presence of alveoli, in whose walls the pulmonary capillaries are distributed and where O_2 and CO_2 exchange occurs.

All the gas exchange airways are approximately the same diameter (0.05 cm) and, like the bronchioles, receive their structural support from the connective tissue framework of lung tissue. There is a circular layer of smooth muscle whose motor innervation is probably similar to that of the bronchioles, although this is not clearly defined. The mucosa is a single layer of epithelial cells in the respiratory bronchiole, some of which contain cilia. There are no goblet cells. A new type of cell, believed to be a serous secretory cell, is present interspersed with the low cuboidal epithelium. Sensory nerves have been described in the mucosal layer. These are undifferentiated types similar to those seen in the bronchioles, but much less frequent. There is no airway epithelium in the alveolar ducts.

The blood supply to the gas exchange airways is via the pulmonary circulation, not the bronchial circulation.

Function

The gas exchange airways serve to distribute the inspired gas by bulk flow within the terminal respiratory units (see later). Ordinarily, these airways do not contribute to airway resistance, nor do they contribute to the anatomic dead space, since they contain alveoli. Because of their blood supply via the pulmonary circulation, the smooth muscle of these units (particularly the respiratory bronchioles in man) may be selectively influenced by agents present in the pulmonary blood. (Chapter 7) These airways dilate and contract with lung volume and account for approximately one-third of the alveolar gas exchange volume (4).

PULMONARY CIRCULATION

This portion of the lung will be discussed only very briefly. Except for the alveolar microcirculation (see later), the pulmonary vasculature is not directly exposed to pollutants entering the airways. However, material in the vascular stream from other sources (intestinal absorption, skin penetration) passes through the pulmonary circulation before being distributed to the body through the systemic blood vessels. (See Chapter 7)

Structure

The entire output of the right ventricle enters the pulmonary artery and is distributed to the lungs through the pulmonary arteries. These branch with the airways down to the respiratory bronchioles. Thus, at all levels the pulmonary arteries are adjacent to the airways, although they do not contribute to their blood supply above the gas exchange airways. Larger pulmonary arteries are of the elastic variety whereas the smaller ones, accompanying the terminal bronchioles, are described as muscular. The pulmonary arteries are richly innervated. The motor innervation is via the sympathetic branch of the autonomic system.

The pulmonary microcirculation consists of an extensive capillary network, which currently is thought of more as a sheet of blood rather than individual channels (9). Pulmonary capillaries are somewhat larger than systemic capillaries. The origin of the capillary network is via short precapillary vessels that branch at right angles from the small pulmonary arteries. Each capillary network is continuous over several alveoli before condensing into the pulmonary venous system (15). The pulmonary capillaries do not appear to be directly innervated. They contain no smooth muscle.

The pulmonary venous system is not closely related to the airways and, in fact, is said to lie as far from airways as is anatomically feasible. The veins drain into the left atrium of the heart. The venous system contains smooth muscle and is well innervated, presumably via the sympathetic branch of the autonomic nervous system.

Function

The pulmonary vascular bed is highly distensible and has a low resistance to blood flow. All of the cardiac output (right ventricular output) passes through the pulmonary vascular bed at a driving pressure which is 1/4 to 1/5 that of the systemic circulation. The average length of the pulmonary capillaries is such that the red cells remain in contact with the gas phase for 1/2 to 1 second, which is sufficient for O_2 and CO_2 exchange.

Because it is a flow-through system, there is no dead space comparable to that of the airways. The motor innervation seems related mainly to regulating the volume of the large vessels. Pulmonary vascular smooth muscle is only weakly sensitive to circulating epinephrine and histamine, but is quite sensitive to certain other circulating agents including 5-hydroxytryptamine (serotonin) and fibrinopeptides.

Because all venous blood passes through the pulmonary microvascular bed before entering the systemic circulation, the pulmonary capillaries act as a filter for removal of embolic material and as a site of inactivation of certain chemical substances in the blood.

It has been suggested, because the pulmonary arteries are adjacent to the airways, that inflammatory airway disease may cause scar tissue to extend and encompass the adjacent pulmonary artery, thereby affecting the distribution of blood flow to the segment distal to that point (16).

When the lung parenchyma is injured, the healing process is accompanied by ingrowth of the bronchial (systemic) circulation. There is no clearly documented evidence that the pulmonary vascular bed has any significant growth or regenerative power in the normal adult.

The smooth muscle of the pulmonary microvascular bed, particularly the small muscular pulmonary arteries, is sensitive to the partial pressure of oxygen in the surrounding airspaces. Breathing low oxygen concentrations stimulates the smooth muscle to contract (17).

The endothelium of the pulmonary microvascular bed forms a substantial component of the nonvascular tissue structure of the alveolar walls (approximately 30% by volume). There are data to suggest that the endothelial cells are more sensitive than other lung cells to damage by toxic substances such as high oxygen (18) and endotoxin (19).

THE TERMINAL RESPIRATORY UNITS

Structure

Beyond the last membranous airway (terminal bronchiole) the airways contain alveoli, in whose walls the pulmonary capillaries course and where the exchange of O_2 and CO_2 occurs. The respiratory bronchiole and its subdivisions constitute the terminal respiratory unit (TRU) (1,3). The basic structure of the lung is that of a very large number (100,000 in man) of TRU's in parallel. Each of these units receives ventilation through a terminal bronchiole and perfusion through a pulmonary artery.

The total volume of gas in the TRU is called the alveolar volume (V_A). At the end of a normal expiration, the volume of gas remaining in all the TRUs is the functional residual capacity (FRC) of the lung. In the normal adult, the FRC is approximately 1/3 of the maximal possible volume that all the TRUs can hold (total lung capacity, TLC).

The V_A is partitioned in the TRU between the true alveoli and the gas exchange airways. The airways account for approximately 35% of the total gas volume.

The solid tissue of the TRU at normal lung volume accounts for less than 10% of the total unit volume, that is, tissue density is less than 0.10. The alveolar walls are 4 - 8 microns thick and disposed in a complex three-dimensional array that has been likened to the lamellae in a solid foam (20). The alveolar walls consist of three major elements: the alveolar epithelium, the interstitium, the pulmonary capillary endothelium.

The principal cellular structures of the alveolar epithelium are the broad pavement cells which cover the main alveolar surface (type 1 cells), smaller more active-appearing cells (corner cells, type 2 cells, granular pneumonocytes) and alveolar macrophages. The granular pneumonocytes contain peculiar dense granules which are thought to be the precursors of alveolar surfactant (see below). The relation of the alveolar macrophage to the other two types of cells is no longer certain and it may come from another source.

Although the interstitium of the alveolar walls appears minimal in electron microscopic pictures, it contains all the structural elements of interstitial connective tissue elsewhere (21), and quantitatively accounts for about 40% of the total non-vascular wall structure in man and several other mammalian species (22). An important point about the alveolar wall interstitium is that the pulmonary lymphatic capillaries do not penetrate into it. Recent evidence suggests that the three-dimensional lamellar structure of the alveolar units, acting as a solid foam, contain the preferential drainage network for fluid, protein, and foreign material (particulates) from the alveolar wall to the lymphatic system (20).

As mentioned earlier, the capillary endothelium accounts for some 30% of the non-blood portion of the alveolar walls, making it a large tissue in its own right.

On the surface of the alveolar epithelium, separating the gas phase from the epithelial cells, is a thin film of fluid on whose surface is distributed a monomolecular phospholipid film (alveolar surfactant). It is believed to be actively secreted by the type 2 alveolar cell. Its removal and fate are uncertain.

Function

Current evidence indicates that the entire TRU expands and contracts proportionately with breathing (23). Alveolar surface area changes with lung volume to the 2/3 power. Over the range of lung volume from FRC to TLC, the shape of the alveoli do not change significantly (24).

In breathing, the gas exchange airways receive the bulk fresh air flow coming down the airways, whereas the alveoli receive the air that was in the gas exchange airways and dead space at the end of the previous expiration. Aerosol particles, in the size range of from 0.3 to 2 microns diameter, do not impact or settle out in the conducting airways and, therefore, reach the respiratory bronchioles and alveolar ducts by bulk flow. These particles are poorly diffusible and mix only by mass ventilation with the gas in the gas exchange airways. Gases and particles less than 0.3 microns diameter diffuse from the gas exchange airways into the alveoli themselves.

It is clear, however, that most particulate matter reaching the TRU will be in highest concentration in the central portion of the unit. Therefore, the anatomic finding that foreign material is concentrated at the center of units of this size or larger may simply reflect this higher concentration-deposition. On the other hand, the normal clearance mechanisms, which tend to proceed in a centripetal fashion, may also contribute to the localization of material centrally in the unit.

Variation of alveolar surface area with breathing causes movement in the surface film, with loss of some surfactant material and replacement by new molecules. It was thought by Macklin (25) that movements of the alveolar surface caused a surface film to flow up and down over alveolar walls to the central airway (first order respiratory bronchiole) of the TRU. At this point, the material plus any entrained foreign matter on the surface was believed to penetrate through the respiratory bronchiole epithelium into the interstitium, where the lymphatic clearance mechanism operated. This no longer appears to be a likely mechanism. Alveolar clearance is probably mainly by macrophage ingestion or by penetration through the alveolar epithelium and clearance through the interstitium.

The fresh air ventilation of the TRU is useful only if it is matched by a proportionate blood flow through the pulmonary capillaries of that unit.. The relationship of ventilation, $\dot{V}$, to blood flow, $\dot{Q}$, for each TRU is denoted by the ventilation-perfusion ratio, $\dot{V}/\dot{Q}$. Summation of the ventilation/perfusion ratios over all of the TRUs of the lung gives the overall ventilation to perfusion ratio of the lung, and determines the arterial O_2 and CO_2 pressures by which the efficiency of the gas exchange process is judged (8).

Within limits, each TRU can regulate $\dot{V}/\dot{Q}$ through changes in respiratory bronchiole and pulmonary artery smooth muscle tone. Disease processes, however, can markedly alter ventilation/perfusion ratios, and are usually manifest by decrease in arterial oxygen pressure or increase in arterial carbon dioxide pressure, that is, decreased efficiency of the gas exchange process.

The alveolar wall cellular elements respond to stressful stimuli, such as infectious agents or particle deposition, but the destructive lesion known as emphysema is described as degeneration with little or no reparative response. Some apparent inability of the lung to repair the damage of emphysema may reside in the three-dimensional geometry of the lung units and in the dynamic forces acting on the alveolar walls with breathing.

The process by which oxygen penetrates from the gas phase of the TRU into the red blood cells in the capillaries is called diffusion. Ordinarily, there is rapid diffusion of O_2 and CO_2 within the gas phase of the TRU. Within the alveolar wall tissues, however, the low solubility of oxygen tends to limit its transport. This is not a serious problem for CO_2. Fortunately, the distance over which oxygen molecules have to diffuse to reach the red blood cells does not exceed half the width of the alveolar walls, that is, 2 - 4 microns. This is readily accomplished in the normal lung, so that the blood leaving the pulmonary capillaries is in equilibrium with the alveolar gas. Under conditions of stress in normal lungs, such as exercise or breathing of air low in oxygen content, the diffusion process may become the limiting factor in gas exchange and the arterial oxygen tension may decrease slightly. In diseased conditions, however, wherein alveolar capillaries are destroyed or the alveolar wall between the red cells and gas phase thickened due to inflammatory responses, the diffusion process may be adequate at rest but not under conditions of stress, which will then lead to a decrease in the arterial oxygen tension. This defect is known in the clinical literature as the alveolar-capillary block syndrome. Even in severe pathologic conditions this is not a serious problem at rest; other associated problems of $\dot{V}/\dot{Q}$ are generally more serious.

The force necessary for lung expansion during breathing (i.e., the pressure difference between the inside and outside of the lung, or transpulmonary pressure) is normally quite small. The distensibility of the lung, as measured by the change in lung volume for each unit change in transpulmonary pressure, is called the lung compliance. In diseases in which the alveolar structure is destroyed, the units may be more easily distended and even remain distended during expiration. In diseases leading to fibrotic changes, the TRU may resist changing volume. They are then said to be stiff and to have a low compliance.

METHODS OF STUDY

Several new methods for studying lung structure and function have been developed. Rapid fixation by perfusion with gluteraldehyde and rapid freezing have contributed much toward visualizing the lung as it is in the living state and have given new insight into the interpretation of physiologic studies of lung function. Quantitative histology and statistical sampling have been applied to the lung, yielding important new information. New microscopic tools such as high resolution autoradiography and scanning electron microscopy will undoubtedly provide new insights into lung structure and function.

REFERENCES

1. Hayek, H. (1960). The Human Lung. Hafner, New York.

2. Krahl, V. E. (1964). Anatomy of the mammalian lung. In: W. Fenn and H. Rahn, (eds.) Handbook of Physiology, Sec. 3, Respiration, 1: 213, Amer. Physiol. Soc., Washington.

3. Staub, N. C. (1963). Interdependence of pulmonary structure and function. Anesthesiol. 24: 831.

4. Weibel, E. (1963). Morphometry of the Human Lung. Academic Press, New York.

5. Comroe, J. H. et al. (1962). The Lung: Clinical Physiology and Pulmonary Function Tests. (2nd ed.) Yearbook, Chicago.

6. Bates, D. V. and Christie, R. V. (1964). Respiratory Function in Disease. Saunders, Philadelphia.

7. Comroe, J. H. (1965). Physiology of Respiration. Yearbook, Chicago.

8. West, J. B. (1965). Ventilation Blood Flow and Gas Exchange. Blackwell, Oxford.

9. Sobin, S. S., Tremes, H. M. and Fung, Y. C. (1970). Morphometric basis of the sheet flow concept of the pulmonary alveolar microcirculation in the cat. Circulat. Res. 26: 397.

10. Widdicombe, J. G. (1963). Regulation of tracheobronchial smooth muscle. Physiol. Rev. 43: 1.

11. Kilburn, K. H. and Salzano, J. V., (eds.). (1966). Symposium on structure, function, and measurement of respiratory cilia. Amer. Rev. Resp. Dis. 93 (Part 2): 1.

12. Reid, L. (1968). Bronchial mucus production in health and disease. In: A. Liebow and D. Smith (eds.). The Lung. Williams and Wilkins, Baltimore, 87.

13. Staub, N. C. (1971). The structural basis of lung function. In: C. Gray and J. Nunn (eds.). General Anesthesia. (3rd ed.). Butterworths, London.

14. Macklem, P. T. and Mead, J. (1967). Resistance of central and peripheral airways measured by a retrograde catheter. J. Appl. Physiol. 22: 394.

15. Staub, N. C. and Schultz, E. L. (1968). Pulmonary capillary length in dog, cat and rabbit. Resp. Physiol. 5: 371.

16. Naeye, R. L. (1968). Pathology of the pulmonary circulation. In: A. Liebow and D. Smith (eds.). The Lung. Williams and Wilkins, Baltimore, 164.

17. Staub, N. C. (1969). Respiration. Ann. Rev. Physiol. 31: 173.

18. Kistler, G. (1967). Development of fine
     structural damage to alveolar and capillary
     lining cells in oxygen-poisoned rat lungs. J.
     Cell Biol. 32: 605.

19. Finegold, M. J. (1967). Interstitial
     pulmonary edema. Lab. Invest. 16: 912.

20. Staub, N. C. (1970). The pathophysiology of
     pulmonary edema. Human Path. 1: 419.

21. Low, F. N. (1961). The extracellular portion
     of the human blood-air barrier and its relation
     to tissue space. Anat. Rec. 139: 105.

22. Weibel, E. (1969). The ultrastructure of the
     alveolar-capillary membrane or barrier. In: A.
     Fishman and H. Hecht (eds.). The Pulmonary
     Circulation and Interstitial Space. Univ. of
     Chicago.

23. Storey, W. F. and Staub, N. C. (1962).
     Ventilation of terminal air units. J. of Appl.
     Physiol. 17: 391.

24. Klingele, T. C. and Staub, N. C. (1970).
     Alveolar shape changes with volume in isolated,
     air-filled lobes of cat lung. J. Appl.
     Physiol. 28: 411.

25. Macklin, C. C. (1955). Pulmonary sumps, dust
     accumulations, alveolar fluid, and lymph
     vessels. Acta Anat. 23: 1.

CHAPTER 3. TRACHEOBRONCHIAL RESPONSE TO INSULT

P. MACKLEM, Respiratory Division, Royal Victoria
Hospital, McGill University, Montreal, Canada

KAYE H. KILBURN, Division of Environmental Medicine,
Duke University Medical Center,
Durham, North Carolina

The various tracheobronchial responses to insult
may be divided into acute and chronic processes:

<u>Acute</u>

Cough
Bronchoconstriction
  reflex
  direct
  humoral
    transmission
Mucus secretion
Ciliary dysfunction
Inflmmation

<u>Chronic</u>

Mucus Secretion
  mucous gland
    hyperplasia
  goblet cell
    metaplasia
Inflammation
Ciliary dysfunction
Neoplasia
Bronchoconstriction (?)

ACUTE RESPONSES

The cough reflex is produced by stimulating
cough receptors which are situated principally in
large airways (1). Exogenous (inhalation of
irritants) and endogenous (failure of mucus clearance
due either to excessive production or deficient
ciliary activity) stimuli initiate cough. The
effectiveness of cough as a clearance mechanism
depends on the linear velocity of the gas (2,3).
This is greatest in those airways which are
dynamically compressed during forced expiration and
has been estimated to approach the speed of sound in
the trachea (4), although direct measurements
recorded linear velocities of only 1/3 of this value

(2). At high lung volumes dynamic compression does not extend beyond lobar or segmental bronchi in normal lungs, so that cough is most effective in clearing the larger airways (2,3).

Bronchoconstriction may be stimulated by reflexes, by direct effects of irritants on the airway smooth muscle and by release of humoral, transmitter substances (5,6). Bronchoconstriction can occur either by an increase in the number of impulses traveling via the parasympathetic nerves or by a decrease of the traffic through sympathetic fibers (allowing the parasympathetic nerves to act unopposed) (7,8,9). In the dog at least, the autonomic nervous system appears to be arranged so that smooth muscle constriction may occur either in small (<2mm. diameter) or larger (>2 mm. diameter) airways or both, but whether or not this serves a useful function in terms of survival is unknown (9). The effects of bronchoconstriction on pulmonary function depend upon not only the degree of constriction, but also the site, and whether or not it is evenly or unevenly distributed along parallel pathways (10). The site of reflex bronchoconstriction and of that due to release of humoral transmitter agents is unknown. Bronchoconstricting agents delivered to the lung via the pulmonary vasculature cause constriction in small airways including alveolar ducts (11). If they are delivered via the bronchial circulation, larger airways constrict (12). Constriction in cartilaginous airways tends to protect them against closure (13). In the non-cartilaginous bronchioles maximal constriction causes closure but it is possible that lesser degrees of smooth muscle tone might protect against closure (10).

Mucus secretion is under autonomic and probably chemical control. Little is known about the factors governing the chemical composition and volume of the mucus that is secreted (14). Mucus is cleared by ciliary activity or, failing that, by the cough mechanism. When mucus is produced in excess, or retained due to ciliary dysfunction, it causes airway obstruction by plugging airways (14), and possibly by replacing the surfactant that probably lines bronchioles (15,16). A wide variety of irritants leads to ciliary dysfunction in vitro and may well cause mucus retention in vivo (14).

Acute inflammation of bronchi and bronchioles can result from viral, bacterial, and chemical agents.

CHRONIC RESPONSES

Goblet cell metaplasia is a non-specific response to any chronic irritation of a mucus membrane (17). It has been produced in the tracheobronchial tree by experimental viral infections, injection of formalin into the airways, exposure to nitrogen dioxide, phosgene, sulfur dioxide, and cigarette smoke (10). It presumably leads to excessive mucus production. The major effects of goblet cell metaplasia may result from the development of these cells in bronchiolar epithelium. Normally, mucus is neither produced nor found in bronchioles. In fact, there is considerable evidence, both direct and indirect, that these airways are lined with surfactant (15,16). Replacement of surfactant with mucus at this level, but not in the alveoli, as might occur with goblet cell metaplasia in bronchioles, should result in airway narrowing and closure leading to an increased flow resistance and gas trapping, both characteristic features of chronic bronchitis (10). As cilia are not found in bronchioles, and because coughing is ineffective in clearing small airways, mucus at this site may be effectively trapped. Looked at in this light, chronic bronchitis may well be a non-specific response to any chronic irritation of the tracheobronchial mucus membrane. Chronic cough and expectoration of sputum may be attributable to the mucus gland hyperplasia, whereas the manifestations of airway obstruction may be attributable to goblet cell metaplasia in bronchioles.

Chronic inflammation of small bronchi and bronchioles is a common morphological finding in patients dying with chronic irreversible airway obstruction (18). Factors which determine whether an acute inflammatory reaction resolves completely or becomes chronic are poorly understood, but retention of secreted mucus due to chronic ciliary dysfunction, impaired cough, and production of mucus in bronchioles could hardly be expected to assist in the resolution of an acute inflammatory reaction. Indeed, mucus plugs observed at post-mortem are occasionally in the process of being organized.

Whether or not this is a common occurrence in chronic bronchitis is unknown, but if so, it could explain the fibrous obliteration of small airways originally postulated by Laennec to account for the airway obstruction (19).

Whether or not "chronic" bronchoconstriction occurs as a tracheobronchial response to injury is undetermined. However, many patients with irreversible or only partially reversible airway obstruction wheeze and are considered to have an "asthmatic" component to their chronic bronchitis. The occasional dramatic improvement of the patient with chronic airway obstruction following therapy with corticosteroids could be due, either to the anti-inflammatory action of these drugs, or to their effect on the antigen-antibody reaction.

Apart from neoplasia, most naturally occurring chronic responses of the tracheobronchial tree to injury probably involve most of the mechanisms described. To dissect the role of each in producing the clinical and pathophysiological manifestations of disease is probably futile at the present time. Chronic inflammation, mucus hypersecretion, ciliary dysfunction, and impaired cough are so closely interrelated that it is difficult to picture one occurring in the absence of the others, and all contribute to the pathogenesis and pathophysiology of chronic irreversible airway obstruction.

NEOPLASIA

Alterations in the differentiation of airway epithelial cells appear as a continuum, from exfoliation due to the influenza virus, through goblet cell replacement of ciliated cells, to squamous metaplasia in which the pseudostratified columnar epithelium is replaced by keratinized layers of skin like squamous cells. Classic studies of the lungs of patients with bronchogenic carcinoma showed many areas of atypical squamous metaplasia with multiple layers of squamous cells, large atypical nuclei in these cells, and interruption of the basement membrane by nests of such cells! Some of these nests are diagnosable as carcinoma in situ (20).

The most useful generalization to emerge from recent experiments is that the combination of carcinogen exposure and a non-specific stimulus to proliferation will cause tumors, while either alone will not do so. Experiments with trachea rings from mouse or human trachea in tissue culture show that exposure to hydrocarbon carcinogens, such as benzo-a-pyrene or benzanthracene, and hydrocarbon with cigarette smoke condensates, produce atypical squamous cells in clusters as early as eight days after exposure, compared to slight reduction in maturation of cells in control tissue (21,22). Vitamin A deficiency enhances these effects. The larger airways of Syrian hamsters are an excellent model for production of tumors using a variety of chemical agents. Chronic irritation by implanted hooks (23) increases the frequency of tumors, and vitamin A deficiency substantially potentiates benzo-a-pyrene in the hamster (24). Recent experiments have shown that exposure to cigarette smoke produces tumors in hamsters (25) and in dogs which smoke cigarettes through tracheostomies (26). Oxidant gas exposure, specifically nitrogen dioxide, has also been associated with an increase in the number of tumors produced in experimental animals by hydrocarbon carcinogens. The mechanism may be by nitrosamines which are converted to more active forms, alkylating diazoalkanes, during metabolism (27).

Much remains unknown about carcinogenesis, but it appears likely that continuous or intermittent cell injury over long periods by physical and chemical agents produces goblet cell hyperplasia, mucous gland hyperplasia, squamous metaplasia, atypical cellular changes, carcinoma in situ, and finally invasive carcinoma. The irregular pattern of carcinoma prevalence in occupational and environmental groups reflects many factors including dose, primary removal, and the presence and persistence of the active molecule of the carcinogen in the cell. Differences in metabolism between individuals due to genetic factors, pre-existing disease, or subclinical injury may be critical, either in detoxification of the responsible agents, or in their formation within cells. Just as it has been difficult to relate cancer in man to one source or one chemical, it is fallacious to expect that removal of single agents will have a measurable effect on cancer prevalence.

REFERENCES

1. Widdicombe, J. G. (1954). Receptors in the trachea and bronchi of the cat. J. Physiol. 123: 71.

2. Macklem, P. T. and Wilson, N. J. (1965). Measurement of intrabronchial pressure in man. J. Appl. Physiol. 20: 653.

3. Mead, J. et al. (1967). Significance of the relationship between lung recoil and maximum expiratory flow. J. Appl. Physiol. 22: 95.

4. Ross, B. B., Gramiak, R. and Rahn, H. (1955). Physical dynamics of the cough mechanism. J. Appl. Physiol. 8: 264.

5. Widdicombe, J. G. (1963). Regulation of tracheo-bronchial smooth muscle. Physiol. Rev. 43: 1.

6. Widdicombe, J. G. (1966). The regulation of bronchial calibre. In: C. G. Caro, (ed.). Advances in Respiratory Physiology, Arnold, London, 48.

7. Olsen, C. R. et al. (1965). Motor control of pulmonary airways studied by nerve stimulation. J. Appl. Physiol. 20: 202.

8. Woolcock, A. J. et al. (1969). Effect of vagal stimulation on central and peripheral airways in dogs. J. Appl. Physiol. 26: 806.

9. Woolcock, A. J. et al. (1969). Influence of autonomic nervous system on airway resistance and elastic recoil. J. Appl. Physiol. 26: 814.

10. Macklem, P. T. (1971). Airway obstruction and collateral ventilation. Physiol. Rev. 51: 368.

11. Colebatch, H. J. H., Olsen, C. R. and Nadel, J. A. (1966). Effect of histamine, serotonin and acetyl choline on the peripheral airways. J. Appl. Physiol. 21: 217.

12. Dekock, M. A. et al. (1966). New method for perfusion of bronchial arteries; histamine bronchoconstriction and apnea. J. Appl. Physiol. 21: 185.

13. Olsen, C. R., Stevens, A. E. and McIlroy, M. B. (1967). Rigidity of trachea and bronchi during muscular constriction. J. Appl. Physiol. 23: 27.

14. Kilburn, K. H. (1967). Cilia and mucus transport as determinants of the response of lungs to air pollutants. Arch. Env. Health 14: 77.

15. Macklem, P. T., Proctor, D. F. and Hogg, J. C. (1969-70). The stability of peripheral airways. Resp. Physiol. 8: 191.

16. Gill, J. and Weibel, E. R. (1971). Extracellular lining of bronchioles after perfusion-fixation of rat lungs for electron microscopy. Anat. Rec. 169: 185.

17. Florey, H., Carleton, H. M. and Wells, A. O. (1932). Mucus secretion in the trachea. Brit. J. Exptl. Pathol. 13: 269.

18. Bignen, J. et al. (1969). Morphometric study in chronic obstructive bronchopulmonary disease. Amer. Rev. Resp. Dis. 99: 669.

19. Laennec, R. T. H. (1834). A treatise on diseases of the chest and on mediate ausculation. (4th ed.) (translated by J. Forbes). Longman's, London.

20. Auerbach, O. et al. (1957). Changes in the bronchial epithelium in relation to smoking and cancer of the lung. New Eng. J. Med. 256: 95.

21. Lasnitzki, I. (1968). The effect of hydrocarbon enriched fraction from cigarette smoke on mouse tracheas grown in vitro. Brit. J. Cancer 22: 105.

22. Lasnitzki, I. (1956). The effect of 3-4 benz-a-pyrene on human fetal lung grown in vitro. Brit. J. Cancer 10: 510.

23. Laskin, S., Kuschner, M. and Drew, R. T. (1969). Studies in pulmonary carcinogenesis. In: M. G. Hanna, P. Nettesheim and J. R. Gilbert (eds.). Inhalation Carcinogenesis. Proc. of Conf. Gatlinburg, Tennessee. Conf. 69001. Natl. Tech. Info. Serv., Springfield, Virginia, 321.

24. Saffiotti, U. et al. (1967). Studies on experimental lung cancer: Inhibitions by vitamin A of the induction of tracheobronchial squamous cell tumors. Cancer 20: 857.

25. Dontenwill, W. (1969). Experimental investigations on the effect of cigarette smoke inhalation on small laboratory animals. In: M. G. Hanna, P. Nettesheim and J. R. Gilbert (eds.). Inhalation Carcinogenesis. Proc. of Conf. Gatlinburg, Tennessee. Conf. 69001. Natl. Tech. Info. Serv., Springfield, Viriginia, 389.

26. Auerbach, O. et al. (1967). Histological changes in bronchial tubes of cigarette smoking dogs. Cancer 20: 2055.

27. Mohr, U. (1970). The effects of diethylnitrosamine in the respiratory system of Syrian golden hamsters. In: P. Nettesheim, M. G. Hanna and J. W. Deatherage (eds.). Proc. of Conf. Gatlinburg, Tennessee. Conf. 700501. Natl. Tech. Info. Serv., Springfield, Virginia, 255.

CHAPTER 4. MODES OF ALVEOLAR RESPONSE TO INSULT

KAYE H. KILBURN, Division of Environmental Medicine,
Duke University Medical Center,
Durham, North Carolina

The alveoli of the lung react to several hundred damaging agents in a few typical ways. These patterns of response develop in three zones: 1) the alveolar surface, 2) cells of five types, and 3) the interstitial spaces. This chapter describes the responses in these three zones of lung to injurious agents, both natural and man-made. The responses to injury include not only damage but repair at the biochemical and morphological levels. Because of the vast number of parallel respiratory units, only a fraction of which are needed except during severe exercise, considerable damage may go unnoticed. Stated another way, part of this large reserve may be eroded insidiously during sequential damage, but the threshold of awareness, usually shortness of breath, may not be reached for months or years. As has been pointed out in Chapter 2, the alveolar barrier is quite different from the bronchial barrier; although they are highly integrated and interact one with the other, the structural and functional features are, in most cases, distinctive. Clearly, the differences fade at the terminal bronchiole where conducting airways end and gas-exchange airways begin. This paper focuses on alveoli, the gas-exchange level, except in those instances in which obstruction to air flow or to alveolar clearance via the airways, or failure of delivery or removal of blood, contribute to alveolar dysfunction or damage.

ORGANIZATION OF THE ALVEOLAR ZONE

The three hundred million alveoli of the adult lung are beyond comprehension or modelling as separate interacting units. The alveoli are organized into 65,000 secondary lobules by incomplete septa of connective tissue, and by central tubes for air supply (terminal bronchioles) and accompanying blood vessels (arterioles). Each secondary lobule is composed of approximately 5,000 alveoli, associated with seven generations of branching airways, which end with branching concertina-like air tubes, called alveolar ducts. Alveolar walls are the partitions which so divide a lung volume of six liters, and augment the surface area in the walls of alveolar ducts, that 300 million alveoli are formed with a total surface area of 70 square meters. Thus, the surface area of a tennis court is contained in the volume of a lung.

How are alveoli organized? In structure they resemble a group of tea cups with open bowls adjacent to compose the alveolar duct. If they were arranged in a line, 170 alveoli would be an inch long. The interstitial connective tissue that lies between capillaries and alveolar cells helps to organize the cells. This interstitial matrix is of mesodermal origin, as are the capillaries which invest alveoli. The connective tissue includes three types of fibers: collagen, reticulin, and elastin.

What five types of cells characterize the alveolar zone of the lung? Two of these, the large alveolar (type II) cell and the pavement, squamous or (type I) alveolar cell, are of endodermal origin, and are continuations of the epithelial lining of the trachea and conducting airways. The other three cell types are the lining cells of blood capillaries (endothelial cells), the macrophages or scavenging mononuclear cells, and the fibroblasts, all of mesodermal origin. The large alveolar cells and the distal secretory cells in the airway produce materials (surfactant), of great importance to the maintenance of membranes which interface with air and liquid in alveoli (Figure 1).

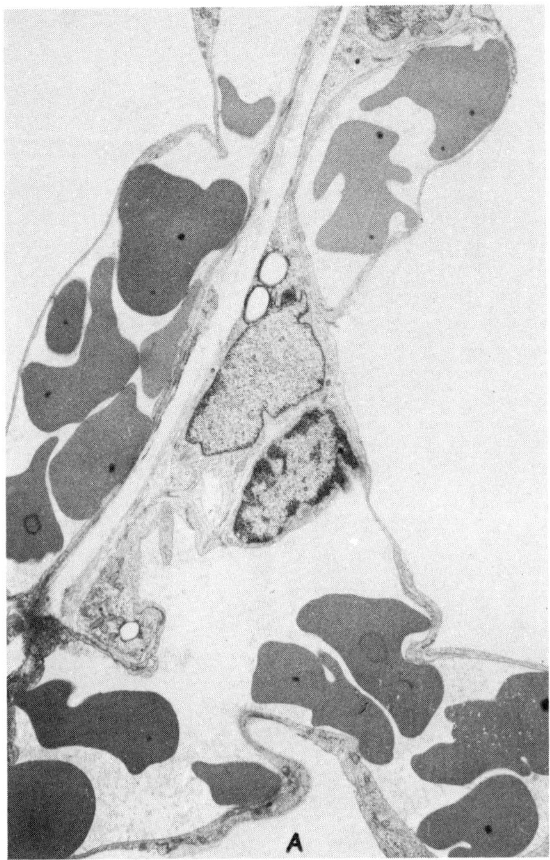

Fig. 1A. Low power electron photomicrograph (1500x) of mammallan (dog) alveolar walls shows capillaries on both sides of connective tissue including two fibrocytes which runs diagonally to the upper right. The thin squamous epithelium of type I alveolar cells covers dark basement membrane. Capillary endothelial cell walls in most capillaries contain darkly staining erythrocytes. The cell nuclei at upper left and below the center are of endothelial cells.

What is the surfactant of the lung? The major surfactant, or surface tension lowering material, is dipalmitoyl lecithin (DPL), which is composed of two fatty acids (palmitic acid) whose carbon atoms are completely saturated with hydrogen, a triglyceride armature, and choline, a B vitamin, bonded through a phosphate linkage. It requires the proximity of protein to reconstitute the crystalline array of phospholipid on the alveolar surface after compression. These protein bonds may be borrowed

41

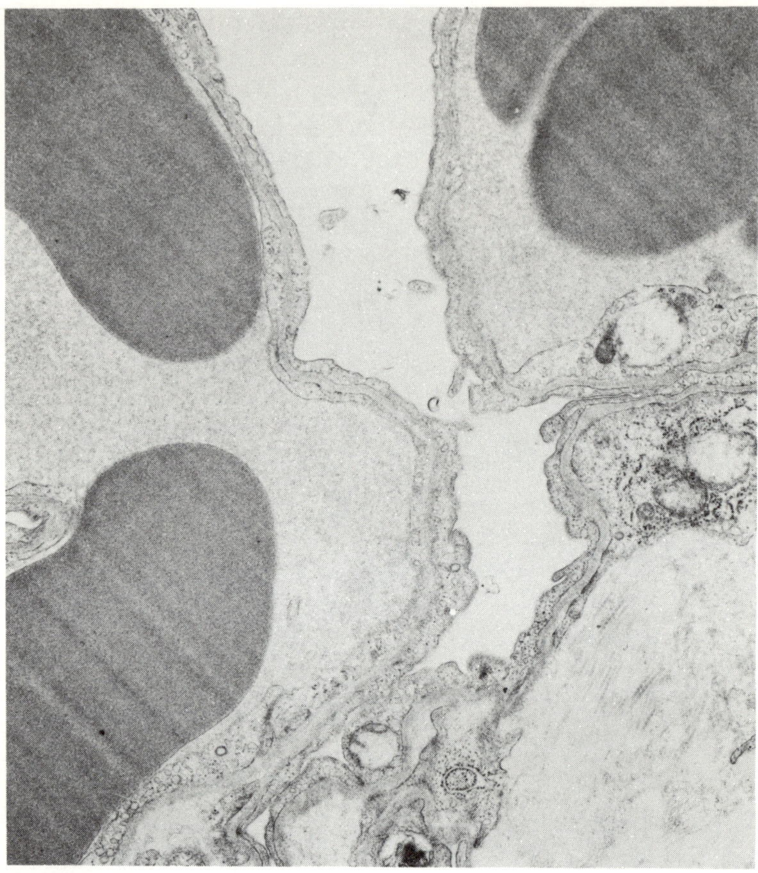

Fig. 1B. Shows details of the layers of alveolar wall from alveolar space in the center at high magnification 5000x. From alveolar space there is epithelial cell, basement membrane, endothelial cell then plasma and dark erythrocytes. (Glutaraldehyde, post fixed with osmium, uranyl acetate, lead citrate.)

from cell membranes or supplied by small or mini-proteins. Studies utilizing radioactive isotopes to label palmitic acid (the saturated fatty acid of DPL) and choline (the B vitamin) show that, when these are injected intravenously, they appear first in the vacuoles of type II alveolar cells and secretory cells of the distal airways, whence they can be delivered very quickly to the cell surface. It is presumed, from analysis of material obtained by careful repeated lavage after the capillaries are blocked, that this material, which has a surface tension of zero, coats alveoli (1).

What does surfactant do on alveolar walls? This coating lowers the surface tension to values less than 10% of that of water or blood plasma, so that the pressure exerted by the curved surfaces of small alveoli does not withdraw fluid from the interstitial spaces between cells or from capillaries. The coating presumably also provides a more easily renewable, and thus economical, protective layer than do cell membranes when exposed to oxygen and more toxic gases (2). It may also strengthen alveolar defenses against the small particles which are carried into alveoli by the air stream, and which have not impacted against the mucus coated surfaces of the conducting airways. Particles coated with DPL may be more easily phagocytized, or may move on the surface layer to the conducting airways, without requiring cellular processing. Alveoli remain dry and stable at low lung volumes by virtue of surfactant, while in its absence the small airways would fill with fluid or collapse.

How do the cells, surfaces, and interstitial spaces react to insults from environmental agents? The obvious and important mode of entry for damaging materials is through the conducting airways. Less obvious is the pathway via the blood which circulates through the lung in greater quantity than through any other organ. Indeed, all of the blood must pass through the lung for oxygenation and removal of carbon dioxide. In a sense, then, the alveoli are in double jeopardy, exposed on one side to all the materials which are not removed from inspired air by the conducting airways, and on the other to materials in the blood. Substances, absorbed from the intestinal tract have, of course, been filtered through the liver before entering the lung, so that certain toxic materials may have been removed. But others, such as certain viruses and chemicals like nitrosamines and paraquat, appear to be destined for the lung even when they are ingested or picked up through the skin. (See Chapter 7)

In the remainder of this chapter the responses of the alveolar surface, the five types of cells, and the connective tissue matrix to insults, particularly from the environment, will be described.

CONTROL OF ALVEOLAR FLUID

There are two means by which the surface properties of the lung could be altered to produce disease in response to environmental agents: first by transudation of fluid, and second by failure of clearance or of reabsorption.

Fluid leaks into alveoli when the pressure in the pulmonary veins is increased by congestive heart failure or by narrowing of the mitral valve between the left atrium which receives the pulmonary blood flow, and the left ventricle which pumps it out to the periphery. Damage to the surface membrane, the epithelial cells or more strikingly the endothelium, by oxidant gases such as nitrogen dioxide or ozone, causes swelling of the cells and leakage of plasma into alveoli (Figure 2). This plasma, containing protein and lipids from plasma and damaged cells, replaces surfactant at the air-liquid interface and disturbs the balance between curvature and surface tension maintained by DPL (3). The sensitivity of the alveolar membranes and cells has not been determined with precision. The doses of oxidant gases required to produce widespread damage in the lung are usually in excess of the levels detected in atmospheric air pollution. However, damage to the endothelial cell membranes of pulmonary capillaries has been demonstrated by electron microscopy in dogs after one hour of exposure to 3-5 ppm of NO_2 (Figure 2) (4), which is near the peak levels of NO_2 found in urban air. Damage may be repaired when exposure is discontinued or even during continued exposure. If large numbers of alveolar capillaries escape severe damage, endothelial proliferation, or membrane repair, restores continuity of the circulation, while the lymphatics of the lung remove the fluid. Lung sections examined with the light microscope show dilation of the lymphatics which surround bronchioles and blood vessels after alveolar damage from NO_2 (4) or hypoxia and in the respiratory distress syndrome (5).

Several materials which make endothelial cells leak may be carried into alveolar capillaries by the circulating blood (6). These include the herbicide paraquat (Weedol) (7), dinitrophenol, thiourea, industrial chemicals such as nickel carbonyl, and

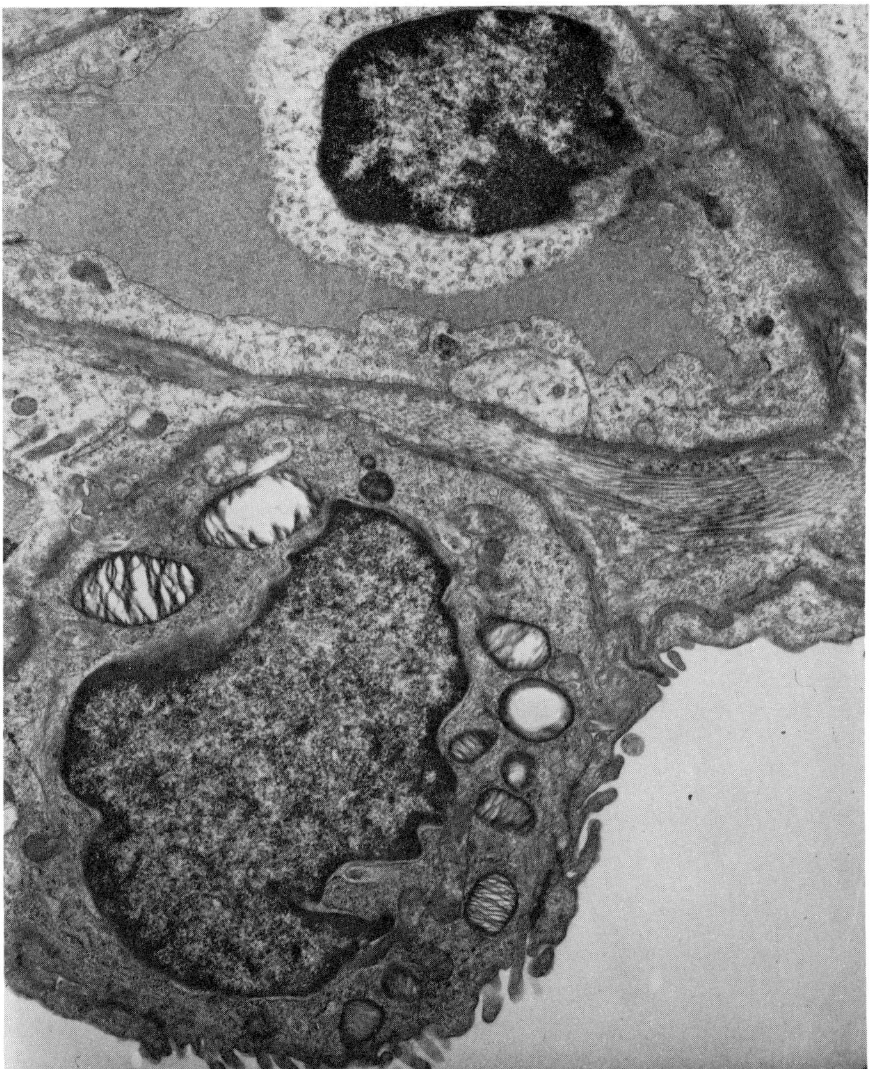

Fig. 2. A higher power electron microscopic picture of a capillary (above) after one hour's exposure of the dog to nitrogen dioxide, 12 parts per million (ppm). Endothelial membranes are vacuolated and pinocytotic vesicles are increased (compare to Fig. 1B). The type II cell below and interstitial space filled with collagen fibrils were not affected. (5000x magnification.)

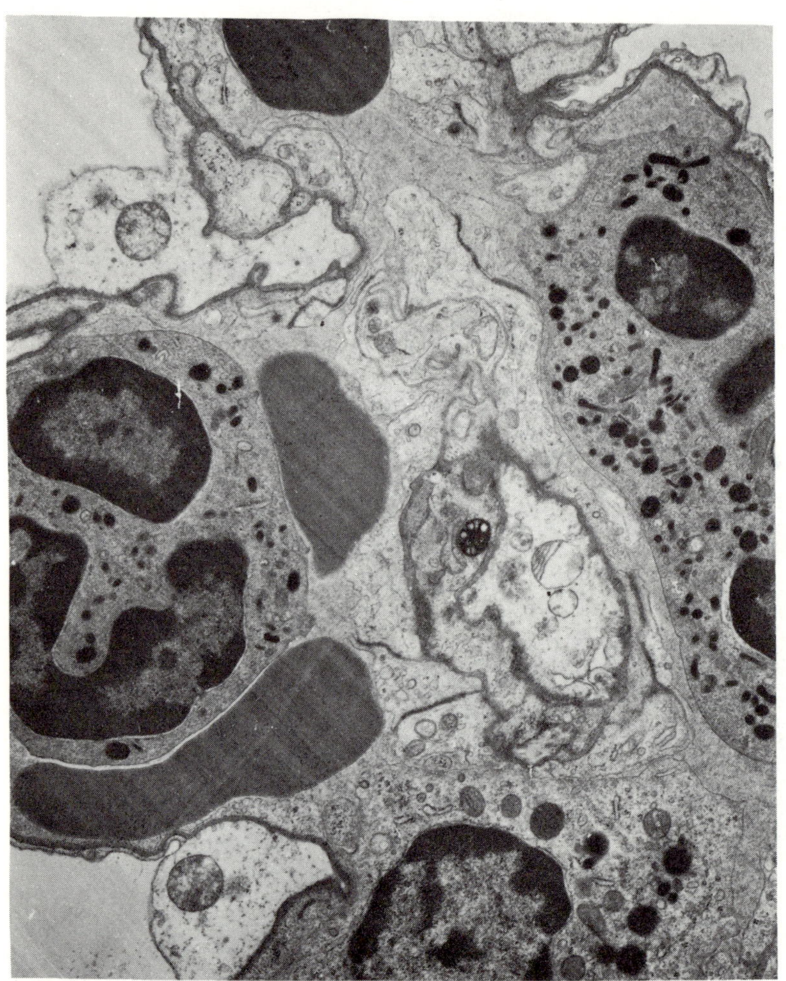

Fig. 3. One day after 13 mg/kg of paraquat was given intraperitoneally. Alveolar epithelial membranes are swollen, into alveolar spaces in upper corners and center of photographs above dark basement membranes. Capillary endothelium is also swollen around erythrocytes and neutrophilic white blood cells in capillaries above, below and on both sides. (5000x magnification.)

cadmium fumes, as well as excess metabolic products such as cholesterol or fatty acids. Physical agents such as heat and ionizing radiation, and trauma such as that produced by explosive decompression or rapid deceleration, also produce pulmonary edema. Paraquat administered intravenously or intraperitoneally to hamsters causes swelling of the membranes and mitochondria of alveolar surface (type I) cells, apparently by shorting out electron transfer in the energy chain (Figure 3). Without energy the sodium pump fails and sodium and water leak into cells; the interstitial space also becomes edematous but endothelial cells are spared. The epithelial cells may be more vulnerable because they are at the same time exposed to the highest alveolar O_2 tensions and further from energy supplies in plasma. The response to paraquat is in sharp contrast with that to oxidant gases (NO_2 and ozone), which damage endothelial cells predominantly. Perhaps a few of the many millions of alveolar capillaries are leaking water, ions and albumin much of the time from damage incidental to the lungs in filtering air and blood (Figure 4).

If the fluid in the alveolar spaces is visualized as the result of a dynamic equilibrium between transudation and reabsorption, it is clear that reduction in reabsorption would be as adverse as excess transudation. This would occur when the right thoracic duct is obstructed so that lymph cannot escape by the usual pathway, or when venous pressure is raised on the left side of the heart as in congestive heart failure.

CLEARANCE FAILURE

The failures of clearance which we must consider are two: upstream failure, or bronchial obstruction, so that the alveolus does not have a normal path of exit up the airway (8); and alveolar failure when, for want of fluid or scavenging cells (macrophages), debris is not cleared from the distal airspaces.

Bronchial obstruction produced by excess quantities of mucus producing cells, resulting from differentiation of the secretory cells in the small airway, clogs the major bronchiolar exit pathway for mucus surfactant, cell debris and cells (9). Within a brief time these alveoli become filled and cease

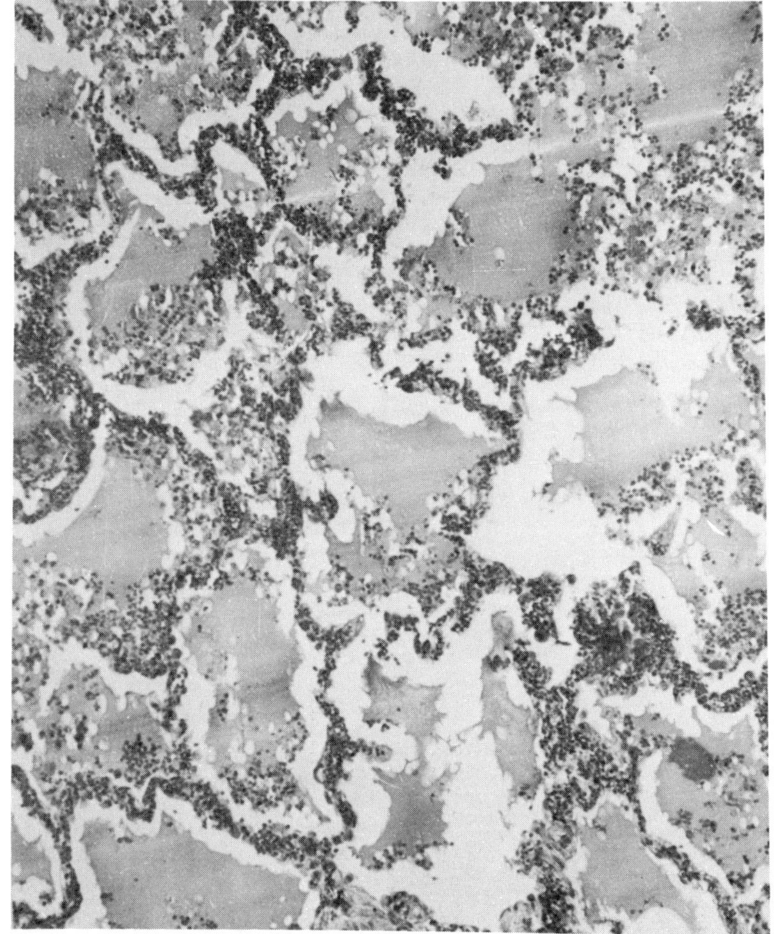

Fig. 4. Pulmonary edema, shown here at low power, occurs when alveolar membranes leak plasma, especially albumin and water, so that alveolar spaces become fluid filled. Some erythrocytes and macrophages are found in the alveoli in the right upper corner. (Hematoxylin-eosin paraffin embedded section of human lung at 25x magnification.)

functioning. If the bronchiole is obliterated or permanently obstructed, the secondary lobule may still receive air via collateral passages from its neighbors, but it rarely functions effectively because clearance via these secondary channels is ineffective.

Alveolar failure may be produced by mucus impaction just described, but two other examples illustrate this process more specifically. These are pulmonary alveolar proteinosis, and cholesterol or lipoid pneumonia. In the first, the removal of surfactant-like materials and cell debris is incomplete, perhaps due to a failure of macrophage recruitment or failure of the macrophages to recognize the altered surfactant as sufficiently foreign to be ingested (10). This produces a "silting up" of the alveoli with phospholipid micelles and large engorged macrophages. The relationship of this process to inhalation or circulation of specific environmental agents is not clear. However, alveolar proteinosis is described as an early stage of alveolar response in silicosis (11).

The so-called cholesterol or lipoid pneumonias differ in that phagocytosis by macrophages appears to be effective, but these cells remain in the alveolar spaces, instead of passing up the airways, so that air spaces are converted into a culture of cells. From rabbit experiments it appears that simply feeding excessive quantities of cholesterol or lipid will produce this picture; but it could also occur because of poisoning of a macrophage enzyme system or by some other interference with macrophage physiology.

Pigment patterns in adult human lungs at autopsy reflect the life-long equilibrium between deposition or synthesis and removal. That these collections are around airways and vessels, or associated with connective tissue septa, including the pleural, suggests that these are poorly cleansed zones (2).

Recruitment of cells into the alveoli and proliferation of the alveolar lining cells can both produce serious functional impairment. If the process is generalized the organism becomes hypoxic and dies.

Such cell recruitment occurs in bacterial pneumonia; for example, that caused by the Pneumococcus. The steps in this process, from delivery of bacteria into the alveoli, through recruitment of blood leucocytes, and subsequent invasion by large numbers of macrophages, are well known (12). Clearly, this constitutes one form of response to environmental agents. Bacteria seem to attract the heterophil or polymorphonuclear white blood cells, while less virulent organisms and antigens attract primarily monocytes. Thus, in tuberculosis, the heterophil stage of response is relatively brief and the alveoli are quickly filled with macrophages, which appear to be descendants of circulating monocytes derived from the bone marrow. A variant of monocytic pneumonia is granuloma formation, which appears to represent the failure of macrophage migration to an area from which they can be removed via the airways. Thus, the cells remain in situ, die, and are phagocytized by their cohorts. Another variation is provided by certain antigens which attract eosinophils in large numbers into the lung. The antigen may be ragweed, a fungus as in farmer's lung (13), or toluene diisocyanate, an industrial chemical used in the formation of polyurethane foam. The last variation of this type of response is endothelial ingrowth, which occurs in unresolved pneumonia where fibroblasts and endothelial cells grow into alveoli without apparent regard for the existing alveolar organization. This has also been observed in patients dying after continuous exposure to high concentrations of oxygen for several days (14).

Proliferation of cells of the alveolar lining can be either benign or neoplastic. It appears that the primary response to generalized damage is replacement of epithelium by the large alveolar (type II) cells. Viruses or chemical damage from cadmium, nickel carbonyl, radiation, phosgene, and nitrogen dioxide, change the cells lining alveoli from the thin attenuated pavement type to cuboidal cells resembling normal type II cells, Figure 5 (15). When the stimulus is brief and self-limited, differentiation and exfoliation occur and type I cells are restored as the alveolar covering. However, if damage continues, or the response cannot be turned off, cuboidal cells remain.

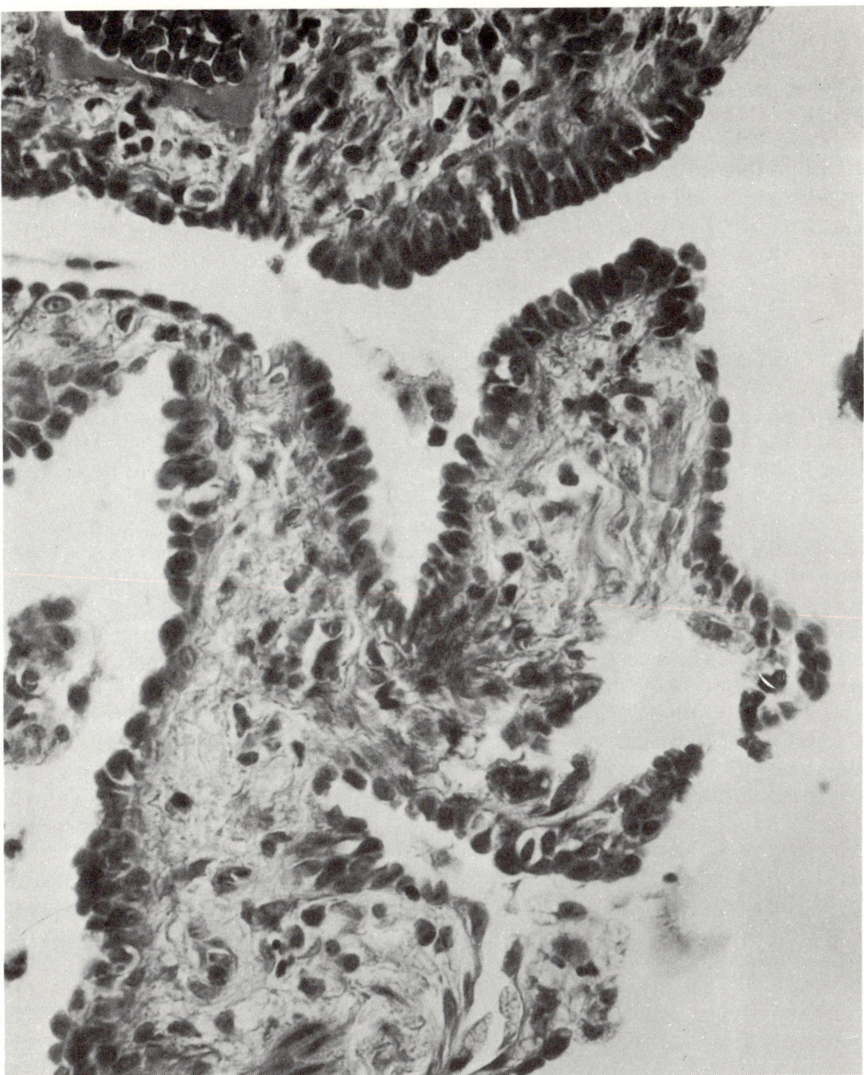

Fig. 5. Cuboidal cells resembling cells of the distal bronchioles or alveolar type II cells line alveoli after damage or after obstruction of airways. Alveolar interstitial spaces are fibrous and thickened and alveoli are obscured or obliterated. (Hematoxylin-eosin, paraffin embedded human lung. 100x magnification.)

Occasionally widespread neoplastic proliferation develops of this cell type, or of the similar neighboring cells in the terminal conducting airways. Alveoli become filled with autonomous cells, which cover the connective tissue matrix and may produce large quantities of lipids or mucin. These materials fill alveolar spaces and are cleared in the sputum. However, clearance is incapable of keeping up when a large portion of the lung is involved and patients die of loss of functioning lung.

INTERSTITIAL CONNECTIVE TISSUE

Disturbances of connective tissue in the lung are of two types: first, an increase of the connective tissue in alveolar walls; second, loss or redistribution of connective tissue so that alveolar duct partitions within the secondary lobules are lost.

Over 150 environmental agents have been identified which are associated with fibrosis, i.e. excessive connective tissue, in the pulmonary alveolar interstitial space. Minerals include silica, asbestos, beryllium, manganese, cadmium, and nickel. Many other agents have been incriminated, including the drugs hexamethonium, hydralazine, and bisulfan; ionizing radiation; and infection. Uremia, cancer, cardiovascular disease, and allergies are often accompanied by pulmonary fibrosis. Diffuse fibrosis is often associated also with the so-called collagen diseases: scleroderma, lupus erythematosus and rheumatoid arthritis. In addition to minerals, chemicals, drugs, and infectious agents, sub-lethal exposure to nitrogen dioxide results in proliferation of the connective tissue matrix (16). The possibility of coincidence exists in some of the associations which have been described. However, the large number of trace agents in the air, plus the numbers of food additives and drugs ingested, together with a large number of instances of pulmonary fibrosis of unknown etiology, suggest that within these coincidences there are causal associations. This highly nonspecific response probably has many ways of being triggered.

Perhaps the most important disease of the alveolar zone results from alterations in the distribution of the connective tissue matrix so that alveolar form is lost. This departitioning or remodelling of alveoli has been recognized for over 100 years and named pulmonary emphysema. At least four varieties are known which have distinct features (17). _Localized emphysema_ is a hole of variable size in the midst of otherwise normal lung. It may be caused by necrotic pneumonia. Bacterial and the cellular enzymes, inadequately cleared, digest alveolar walls. The bacilli common in hospital acquired infections, including species of Staphylococcus, Pseudomonas, Klebsiella and E. coli, are probably responsible.

Because the lung must fill the thoracic space despite disease processes which cause loss of volume of lung units, several diseases result in scars surrounded by alveolar departitioning and overinflation. Perhaps the best example is tuberculosis, in which healing of granulomas by fibrosis produces contraction. When this is distributed throughout a segment or lobe of the lung, competing contraction pulls upon the remaining lung tissue, producing _pericicatricial emphysema_ around the scars.

The least common and perhaps the most interesting variety of emphysema is distributed uniformly through the entire lung, and results in alveoli of larger size and a correspondingly increased lung volume. It is called _panlobular or panacinar emphysema_. It has been associated with an early onset of breathlessness and often occurs in families, including the women. The relation of this disease to environmental materials is unknown. However, the uniform and generalized departitioning of alveoli, and changes in the interstitial connective tissue of alveoli, suggest an even distribution of an intense level of damage.

The most important type is _centrilobular emphysema_, so named because of its characteristic localization in the center of the secondary lobule, with departitioning of the respiratory bronchiole and its alveolar ducts. This distribution is by no means exclusive; it may select the connective tissue septa, or the subpleural area, or be scattered in all of these areas. While attempts to model localized emphysema and pericicatricial emphysema have been successful, and have shown that the bacterial pneumonia or scattered damage may produce one or the other, efforts to produce centrilobular emphysema failed until recently. It was found that instillation of papain (a vegetable protease enzyme) into the lung (18), or its aerosol administration, produced widespread departitioning, which was centrilobular in distribution, within eight days and often by four days (Figure 6) (19). This suggested that the material was being distributed to, or at least acted predominantly in, the centers of lobules, to digest collagen and other connective tissue elements of alveolar walls. It is interesting that repeated doses of papain, particularly by aerosol, produce a uniform generalized departitioning resembling panlobular emphysema. Attempts to show a change in the quantity of collagen in the lungs of papain treated animals showed instead that whole lung collagen content was unchanged. Although the specificity and sensitivity of these methods are questionable, estimates from morphology that only 10% of the lung's collagen is in alveolar walls, as compared to 90% around vessels, bronchi, and in septa, suggest that important changes are beneath detection by this approach. However, papain stimulated cells to fill with rough endoplasmic reticulum, and the lung slices from these animals incorporated more lysine and proline as measures of protein and collagen synthesis than did controls. It seems justified to make some speculations relating proteolytic enzymes similar to papain which are contained in the lysosomes of many cells and typical alveolar responses to injury.

The role of collections of leucocytes and macrophages in alveoli stimulated by air pollutants such as cigarette smoke is not understood. They may act within the lung to digest connective tissue directly and thus remodel alveoli, or they may activate enzymes within the lung. Alternatively, macrophages may contain or attract antienzymes to neutralize enzymes. The findings that cigarette smokers have more blood leucocytes and macrophages in their lungs than non-smokers (20), and that the instillation of leucocyte preparations destroys alveoli (21), suggest that proteolytic enzymes and

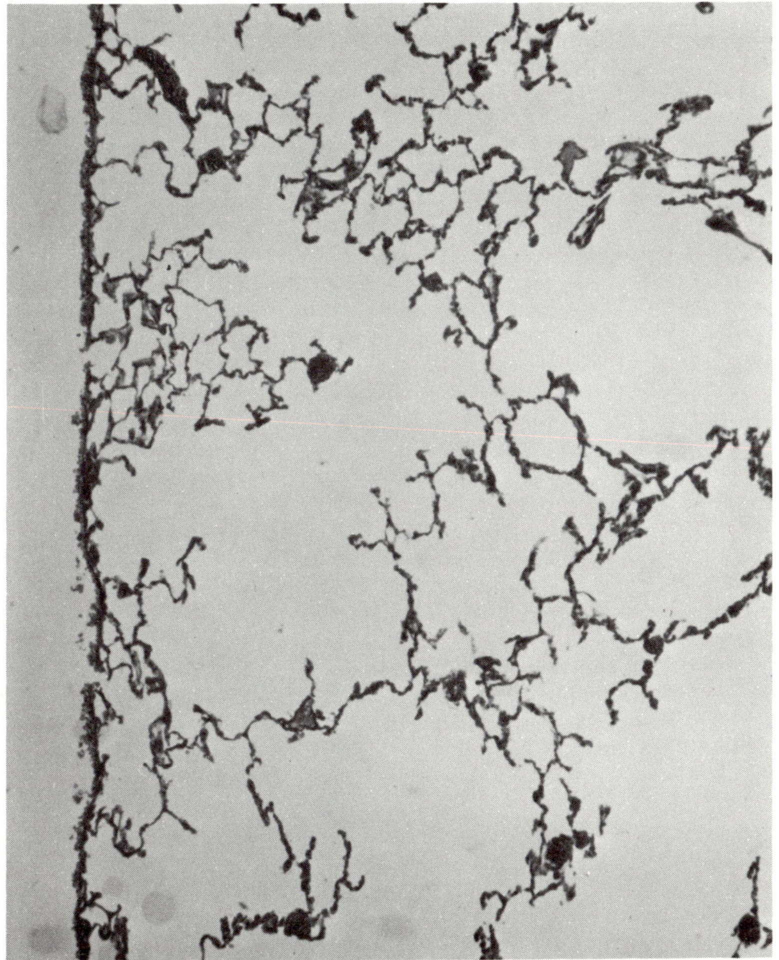

Fig. 6A. The alveoli beneath the pleura of this hamster eight days after 500 µg of papain was given into the lung via the trachea are extensively departitioned.

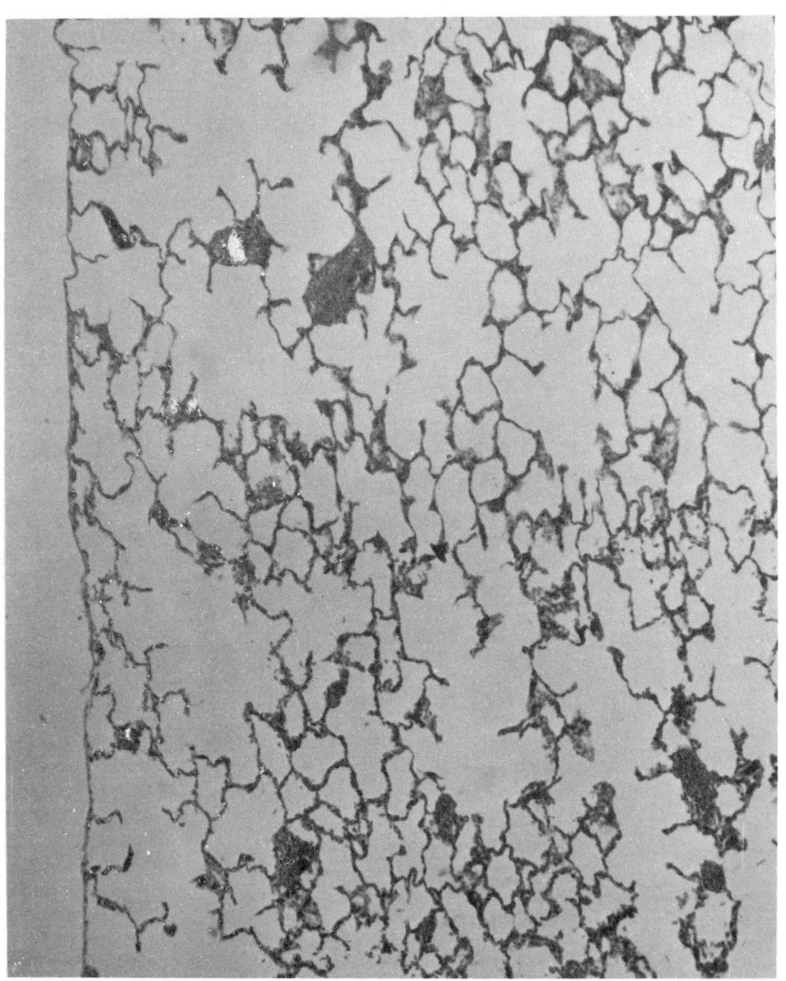

Fig. 6B. Compare the normal lung of a cohort animal which received only saline eight days before. A few alveolar ducts extend near the pleura but the contrast between normal (b) and papain treated is impressive. (Hematoxylin-eosin hamster. 25x magnification.)

their antienzymes may be important (22). Comparison of the pathological picture in a large number of lungs prepared in inflation, with the smoking histories of the patients, has produced strong correlations between emphysema and prolonged cigarette smoking (23). In the light of these studies of human lungs and the experiments with leucocytes, it is difficult to avoid the suggestion that recruitment of cells in large quantities into the alveoli of the lung, and their continuous stimulation by particles and gases, may produce lysis of the cells and digestion of the connective tissue of the lung. It is a curious fact that no lesions which could be early stages of emphysema or cell collections have been identified in human lungs. Perhaps enzymes are discharged into the extracellular lung water and concentrated in the interstitial space and the lymphatics of the centrilobular areas where destruction first occurs. Further studies are needed of these possible mechanisms.

One important tentative conclusion is that the time course of injury and repair is hours (for NO_2 on endothelium) or days (for papain on interstitial connective tissue) rather than years. It is plausible, in the light of the large number of respiratory units arranged in parallel for receiving air and blood, to suggest that injury to the connective tissue of a few units at a time could produce chronic pulmonary insufficiency, including fibrosis and emphysema, after many sequences of damage.

REFERENCES

1. Hurst, D. J., Kilburn, K. H. and Lynn, W. S. (1971). Isolation and surface activity of alveolar lining layer. J. Appl. Physiol. (In press).

2. Macklin, C. C. (1955). Pulmonary sumps, dust accumulations, alveolar fluid and lymph vessels. Acta Anat. 23: 1.

3. Pattle, R. E. (1958). Properties, function and origin of the alveolar lining layer. Proc. Roy. Soc. (London) Ser. B 148: 217.

4. Dowell, A. R., Kilburn, K. H. and Pratt, P. C. (1971). Effects of acute NO_2 exposure on lung ultrastructure, compliance, and the surfactant system. Arch. Int. Med. 128: 74.

5. Lauweryns, J. J. (1970). Hyaline membrane disease in newborn infants. Human Path. 1: 175.

6. Hawkett, R. L. and Sunderman, F. W. (1968). Pulmonary alveolar reaction to nickel carbonyl. Arch. Env. Health 16: 349.

7. Kilburn, K. H. (1971). Ultrastructural signs of alveolar injury by paraquat. Lab. Invest. (submitted)

8. Kilburn, K. H. (1967). Cilia and mucus transport as determinants of the response of lung to air pollutants. Arch. Env. Health 14: 77.

9. Simpson, T., Heard, B. F. and Laws, J. W. (1963). Severe irreversible airways obstruction without emphysema. Thorax 18. 361.

10. Larson, R. K. and Gordinier, R. (1965). Pulmonary alveolar proteinosis: report of six cases, review of the literature and formulation of a new theory. Ann. Int. Med. 62: 292.

11. Corrin, B. and King, E. (1970). Pathogenesis of experimental pulmonary alveolar proteinosis. Thorax 25: 230.

12. Wood, W. B. (1951). White blood cells vs. bacteria. Sc. Amer. 184: 20.

13. Pepys, J. (1969). Hypersensitivity Diseases of the Lung Due to Fungi and Organic Dusts. Karger Company, New York.

14. Pratt, P. C. (1958). Pulmonary capillary proliferation induced by oxygen inhalation. Amer. J. Path 34: 1033.

15. Liebow, A. A., Steer, A. and Billingsley, J. G. (1965). Desquamative interstitial pneumonia. Amer. J. Med. 39: 369.

16. Stephens, R. J. et al. (1971). Ultrastructural changes in connective tissue in lungs of rats exposed to NO_2. Arch. Int. Med. 127: 873.

17. Pratt, P. C. and Kilburn, K. H. (1970). A modern concept of the emphysemas based on correlations of structure and function. Human Path. 1: 443.

18. Gross, P. E. et al. (1965). Experimental emphysema: its production with papain in normal and silicotic rats. Arch. Env. Health 11: 50.

19. Kilburn, K. H., Dowell, A. R. and Pratt, P. C. (1971). Morphological and biochemical assessment of papain emphysema. Arch. Int. Med. 127: 884.

20. Pratt, S. A. et al. (1969). A comparison of alveolar macrophages and pulmonary surfactant (?) obtained from the lungs of human smokers and nonsmokers by endobronchial lavage. Anat. Rec. 163: 497.

21. Moss, B. et al. (1971). Experimental emphysema in dogs: induction by leucocyte homogenates. Fed. Proc. 30: 293.

22. Cohn, Z. A. and Wiemer, E. (1963). The particulate hydrolases of macrophages, II. Biochemical and morphological response to particle ingestion. J. Exptl. Med. 118: 1009.

23. Pratt, P. C. and Kilburn, K. H. (1971). Extent of pigmentation in autopsied human lungs as an indicator of particulate environmental air pollution. In: C. N. Davies (ed.). Inhaled Particles. (In press).

CHAPTER 5. THE EFFECTS OF CHRONIC RESPIRATORY DISEASE ON THE FUNCTION OF THE LUNGS AND HEART

G. F. FILLEY, University of Colorado School
of Medicine, and Webb-Waring Lung Institute,
Denver, Colorado

However chronic bronchitis, emphysema, and pulmonary fibrosis are caused (by environmental pollutants self-administered or otherwise, infectious agents, etc.) these diseases cause shortness of breath and disability by producing four functional disturbances. These are: 1) inability to inspire deeply, 2) inability to exhale freely, 3) maldistribution of air and blood in the lungs, and 4) impairment of oxygen diffusion from air to blood in the lung capillaries. Because these disturbances of lung and circulatory function are not easily detectable by clinical examination, four types of pulmonary function tests (vital capacity, expiratory airflow, arterial blood gas, and diffusion measurements) have proven to be useful in detecting and quantitating these disturbances (Table 1).

Exactly how these disturbances of function cause the symptom of shortness of breath is not known, because of the subjective nature of this symptom. It is known that the mechanical (and often measurable) disturbances which interfere with stretching the lungs and with moving air through the bronchial tree, can be extremely disabling even though the heart and circulation receive and deliver adequate oxygen. Respiratory diseases kill, however, by so damaging the lungs and circulation that their primary function -- O_2 and CO_2 exchange -- is inadequate to support life, a condition properly termed respiratory failure (1).

TABLE 1 -- MEASUREMENTS USED IN RESPIRATORY PHYSIOLOGY

Vital Capacity (VC)	Max. vol. that can be expired after full insp.
Maximum Breathing Capacity (MBC) or Max. Voluntary Ventilation (MVV)	Total expired flow during 15 sec. of vigorous hyper-ventilation reported in li/min.
Forced Expiratory Volume (FEV_1)	Fast (1 sec.) VC = liters of flow in 1st sec.
Max. Mid-Expiratory Flow (MMEF)	Flow rate of mid-half of fast VC; reported in li/sec.
Compliance of Lung ($\Delta V / \Delta P$)	ΔV = change in lung volume or depth of one breath; ΔP = intrapleural pressure change with breathing.
Resistance to gas flow ($\Delta P / \dot{V}$)	ΔP = alveolar-to-mouth gas pressure difference; $\dot{V}$ = total flow rate.
O_2 Uptake ($\dot{V}O_2$)	Depends on metabolic rate and blood flow.
CO_2 Output ($\dot{V}CO_2$)	Depends on metabolic rate and ventilation.
CO Uptake ($\dot{V}CO$)	Depends on diffusing capacity and ventilation.
Ventilation-perfusion Ratio (V/Q)	Usually refers to regional gas and blood flows.
Alveolar CO_2 Tension (P_ACO_2)	Usually equals arterial CO_2 tension, P_aCO_2.
Alveolar O_2 Tension (P_AO_2)	$P_AO_2 = (P_B - 47) - P_ACO_2 - P_AN_2$.
Arterial O_2 Tension (P_aO_2)	Depends on Hb dissociation curve and lung function.
Aa PO_2 Difference	$P_AO_2 - P_aO_2$.
Diffusing Capacity (D_LCO)	$\dot{V}CO / P_ACO$; also called transfer factor.

The four main pathophysiological disturbances of chronic lung disease will now be described.

RESTRICTION OF INSPIRATION

The inspiratory force required to stretch the lungs is normally expended in overcoming the alveolar surface tension and the elastic recoil forces of the lungs and chest wall as air gains access to the alveoli -- normally to virtually every alveolus. Three types of pathological change can render the lungs hard to stretch: a) when lung surface tension is high, as in certain premature infants; b) when abnormal (e.g. scar) tissue replaces normal tissue; or c) most commonly, when incomplete or non-uniform airway obstruction or fluid accumulations in the lung prevent air from gaining access to all alveoli. For practical purposes the test reflecting this failure of the lungs to be inflated easily is the vital capacity (VC); for laboratory purposes the determination of choice is the lung compliance. The VC is easy to measure and widely applicable in clinical and preventive medicine, but fails per se to pin point which particular pathological change has occurred; determining the lung compliance is difficult, but is occasionally necessary for accurate diagnosis.

Although a low vital capacity often reflects restriction of inspiration, it should not be interpreted without other evidence to indicate "restrictive lung disease" or the laying down of pulmonary scar tissue. For example, asthma, which does not scar the lung, limits inspiration mainly by interfering with the previous expiration; asthma thus distends the lungs (with trapped air) so that there is inadequate room for a deep inspiration, thus lowering VC (2). Neither lung distension (without lung destruction) nor a low VC per se affect gas exchange significantly; most asthmatic patients have normal arterial blood oxygen and carbon dioxide because their total ventilation per minute and ability to distribute air and blood in the lungs are usually not seriously impaired.

AIRWAY OBSTRUCTION

Dyspnea (i.e. shortness of breath or breathlessness) is more commonly caused by airway

obstruction than by any other lung disturbance. In addition to interfering with air entry, obstruction of the airways partially blocks and hence slows the flow of air out of the lungs even more, especially in emphysema, a disease in which lung destruction (in addition to distention) is associated with airway collapse during expiration. Such collapse greatly limits the maximum breathing capacity (MBC or MVV, a useful test of overall ventilatory function). Chronic bronchitis also causes expiratory obstruction, but how it does so is not clear. Excessive mucus production, thickening of mucous membranes by infection and edema, bronchial muscle hyperreactivity and small airway distortion may all contribute to a slowing of expiratory airflow. Such slowing occurs to some extent even in early cases of chronic respiratory disease and mild forms of asthma. For this reason tests which specifically measure expiratory air flow (such as FEV_1 and MMEF) are extremely helpful, not only in evaluating the commonest respiratory symptom (dyspnea), but in detecting the commonest disabling lung diseases in their early stages (chronic bronchitis and emphysema).

Tests of airflow should be carried out by methods which provide a permanent record, so that their validity can be examined. This is becoming more and more necessary as more and more patients and potential patients are being tested by technical personnel in survey studies. A recording spirometer provides a reasonably sensitive and reliable instrument; more sensitive methods require the use of body plethysmography (3) and special electronic means of estimating airway resistance. Some of these methods are so sensitive that they can detect the transient airway obstruction produced by smoking one cigarette (4). Sensitive tests of function must obviously be used with caution in diagnosing disease.

A very important consequence of airway obstruction is carbon dioxide retention. The increased work of breathing (needed to overcome high airway resistance) both limits and leads to a compensatory lowering of alveolar ventilation, so that CO_2 excretion lags behind CO_2 production. Chronic retention of CO_2, though not disabling in itself, encroaches on acid-base reserve mechanisms, lowers blood pH and portends eventual failure of the oxygen transport system including the heart. Retention of CO_2 is best diagnosed by arterial blood gas analysis and is accurately measured by the CO_2 pressure (P_aCO_2) in this blood.

MALDISTRIBUTION OF AIR AND BLOOD IN THE LUNGS

Chronic respiratory disease does not affect all of the hundreds of thousands of bronchial tubes and the 300 million alveoli in the human lung identically; some tubes are more obstructed than others; some alveoli are more waterlogged or scarred than others. Given a fixed total ventilation, this must mean that certain alveoli receive less fresh air than normal, i.e. they are hypoventilated. Unless the total alveolar ventilation is increased, or the amount of blood flowing through the capillaries of these alveoli is reduced in exactly the right proportion to the amount of local hypoventilation, it necessarily follows that arterial blood will be subnormally oxygenated (hypoxemic). When hypoxemia is severe, the patient's finger tips and lips will be blue (cyanotic). This particular form of air-blood maldistribution (local hypoventilation) is the commonest cause of arterial hypoxemia (often called physiological shunting or V/Q disturbance).

The reason why local hypoventilation is so likely to lead to arterial hypoxemia is physicochemical. "Shunting" with respect to oxygen depends on the fact that the distribution of air in the lungs directly affects only the local pressures of oxygen in the gas phase, and not the quantities of oxygen in the blood capillaries, because of the properties of hemoglobin (5,6). "Shunting" also occurs with respect to CO_2 as well (7) but hyperventilation can easily overcome the effect so that CO_2 retention is rarely severe as a direct consequence of maldistribution of air and blood in the lungs. However, as discussed under "airway obstruction", a high airway resistance limits total alveolar ventilation; hence blood CO_2 pressure (PCO_2) often rises in chronic obstructive disease as O_2 pressure (PO_2) falls.

The heart is secondarily damaged by chronic respiratory disease (especially by chronic bronchitis) as follows. Low oxygen pressure in the lungs acting on pulmonary blood vessels (alone or in combination with the often lowered blood pH) raises the pulmonary artery blood pressure. This pulmonary hypertension forces the right ventricle to increase its work. The increase leads to thickening and eventually to dilatation and failure of the right ventricle, a condition often called cor pulmonale. Loss of pulmonary vessels by destructive processes such as emphysema can obviously lead to pulmonary hypertension, but here also hypoxia and low pH are important contributing factors (8), and must be dealt with in treatment.

There is no simple specific test for quantitating V/Q disturbances. However, consideration of the oxygen and carbon dioxide pressures in arterial blood (P_aO_2 and P_aCO_2) in relation to airflow tests can distinguish between hypoxemia primarily due to local hypoventilation, and that caused by inadequate total ventilation, and so provide essential guidance in treating both the pulmonary and the cardiac aspects of respiratory failure.

IMPAIRMENT OF OXYGEN DIFFUSION

The wide variety of diseases which diffusely affect gas exchanging (alveolar) rather than gas conducting (bronchial) tissue can be labeled as "interstitial lung disease." Typically these diseases: a) lower vital capacity because the alveoli are stiffened and lose their normal compliance; and b) lower pulmonary diffusing capacity because the alveolar walls are thickened and lose their normal permeability. If interstitial disease can be established as the cause of a lowered vital capacity, the term "restrictive" lung disease is appropriate. If fibroblastic proliferation (lung scarring) can be established, the term "interstitial fibrosis" applies. If alveolar thickening can be established as the cause of hypoxemia, the term "alveolar capillary block" is appropriate, although it implies an oversimplified picture, i.e. a uniform thickening of alveolar walls uncomplicated by, for example, V/Q disturbances.

The thin (0.1 - 1.0 micron) layer of fluid and tissue interposed between lung gas and capillary blood becomes thickened by many responses to environmental agents. These responses include inflammatory outpourings of pus cells, pulmonary edema (both on the surface of alveolar epithelial cells and in the interstitial space between these cells and the capillaries), overgrowth of epithelial cells and the laying down of scar tissue. Oxygen, being much less soluble in tissue fluids than carbon dioxide, diffuses with greater difficulty through thickened alveolar tissue than does CO_2. Hypoxemia without CO_2 retention is, therefore, commonly found in these interstitial diseases.

As discussed in the previous section, V/Q disturbances can• also produce hypoxemia without CO_2 retention. Testing the lung's ability to take up trace amounts of carbon monoxide is a useful aid in diagnosing diffusion disturbances, because CO uptake by the lung is almost wholly diffusion-dependent. Because CO uptake is somewhat affected by V/Q disturbances, and because these are so common in obstructive disease, a lowered diffusing capacity for carbon monoxide (D_LCO or transfer factor), is most reliable as an indicator of diffusion disturbances in patients without airway obstruction (9).

In summary, the effects of chronic respiratory disease on the lungs and heart are subjective (chiefly manifest as dyspnea), objective (measurable by pulmonary function tests), disabling (by interfering with the capacity for physical exercise and hence many kinds of employment) and eventually fatal (in the case of the most serious types -- emphysema, chronic bronchitis, and irreversible types of interstitial fibrosis). An especially practical guide for the quantitative evaluation of respiratory impairment has been published (10).

REFERENCES

1. Filley, G. F. (1967). Pulmonary Insufficiency and Respiratory Failure. Lea and Febiger, Philadelphia.

2. Bates, D. V., Macklem, P. T. and Christie, R. V. (1971). Respiratory Function in Disease. (2nd ed.) W. B. Saunders Company, Philadelphia.

3. Comroe, J. H. (1962). The Lung. (2nd ed.) Year Book Publishers, Chicago.

4. Nadel, J. A. and Comroe, J. H. (1961). Acute effects of inhalation of cigarette smoke on airway conductance. J. Appl. Physiol. 16: 713.

5. Rahn, H. and Farhi, L. E. (1964). Ventilation, perfusion, and the $\dot{V}_A/\dot{Q}$ concept. In: Handbook of Physiology, Sec. 3, Respiration 1, Amer. Physiol. Soc., Washington.

6. Filley, G. F. (1971). Acid-Base and Blood Gas Regulation. Lea and Febiger, Philadelphia.

7. West, J. B. (1965). Ventilation, Blood Flow and Gas Exchange. Blackwell, Oxford.

8. Harvey, R. M., Enson, Y. and Ferrer, M. E. (1971). A reconsideration of the origins of pulmonary hypertension. Chest 59: 82.

9. Forster, R. E. (1965). Interpretation of measurements of pulmonary diffusing capacity. In: Handbook of Physiology, Sec. 3, Respiration, 2, Amer. Physiol. Soc., Washington.

10. Gaensler, E. A. and Wright, G. W. (1966). Evaluation of respiratory impairment. Arch. Env. Health 12: 146.

ovit PART II

LUNGS RESPONSES TO
ENVIRONMENTAL FACTORS

CHAPTER 6. AIRBORNE CONTAMINANTS

PAUL E. MORROW, Department of Radiation Biology
and Biophysics, University of Rochester School of
Medicine and Dentistry, Rochester, New York

SOURCES, NATURE, AND RELATION TO RESPIRATORY EFFECTS

The important (or potentially important) air
contaminants can be described under several headings:
(1) Urban Pollution -- This is used to describe
pollution which is common to population centers and
especially those areas which are technologically
advanced; (2) Occupational Pollution -- This connotes
contaminants which are specifically related to
industry, agriculture, and various trades, and which
are not covered by urban pollution; (3) Natural
Pollution -- This pertains to natural sources of
airborne gases and particles which have health
significance; and (4) Personal Pollution -- This
refers to household and personal sources of air
contaminants (Chapter 1).

These classes of air pollutants are coexistent
as a rule. This classification serves mainly to
differentiate the principal sources and kinds of
airborne contaminants and to facilitate their
discussion.

URBAN POLLUTION

There has been a remarkable number and variety
of studies made, on equally diverse groups of
subjects, into the health-related effects of airborne
contaminants in urban areas. The first of the
investigative approaches depended upon a
retrospective examination of episodic pollution in
several specific areas about the world. By this
method, conditions were analyzed in terms of

excessive human morbidity and mortality before, during, and after these acute events (1-3).

The second approach is concerned with both prospective and retrospective studies of urban populations and subpopulations. In general, these have relied upon fairly long-term evaluations of high and low pollution either on an interurban or intraurban basis (4-11). Attempts to develop adequate epidemiological cohorts and control groups have been variable in these studies, but generally they have not been very satisfactory (Chapter 8). Collectively, most of the epidemiological efforts have produced a body of circumstantial information which links excessive urban air pollution to adverse health effects, and which reveals a scale of susceptibility within populations, wherein the chronically infirm, especially those with cardiopulmonary disease, the aged, and the very young are the most conspicuous. These population studies do not provide any proof of causality nor do they indicate the role of specific air contaminants. The available evidence suggests that air pollution effects cannot be simply isolated from concomitant meteorological conditions, the effects of superposed viral and bacterial respiratory infections, or from a group of ill-defined factors which may be called "urban factors".

The third body of information is derived from human research (12-14). This approach, which has utilized normal and susceptible subjects, generally compares, on a short-term basis, the effects of natural high air pollution with those of "clean air". In some cases synthetic smog or specific air pollutants have been examined. These studies, too, support the concept of susceptible individuals, but the temporary induction of functional deficits or the aggravation of symptoms is used as the health-effects criteria. On the whole, these studies underline the epidemiological approach. In the case of the oxidant gas, ozone, and the asphyxiant gas, carbon monoxide, they infer that a close relationship might exist between these agents and the stress experienced by cardiopulmonary patients during episodic or excessive air pollution.

Finally, laboratory studies in animals (15-18) and simple cytological and biochemical systems

(19-20) have provided data which support the human studies in general terms. However, most of the studies have been concerned with normal adults and specific pollutants -- a combination which is typically insensitive unless excessive levels of air contaminants are utilized. The long-term studies of animal populations appear to be most readily interpretable, such as in the susceptibility of the aged.

It is evident from the foregoing studies that urban air pollution and smog are extraordinarily complex and variable mixtures. The major sources of urban pollution are automobiles and municipal, industrial, and residential heat and power plants (stationary sources). Moreover, there are observations which suggest that important interactions or synergisms occur between different contaminants, but the nature of these interactions is largely unknown. The following account of specific urban air contaminants, of necessity, does not consider this important area but simply describes pertinent data on the more pervasive agents or classes of agents.

PARTICULATE MATTER -- Many environmental scientists believe that, while no single material can be considered responsible for adverse health effects of urban population, particulate matter is the most important fraction being measured, because of its relatively high intrinsic toxicity, its vectorial or synergistic roles with other pollutants, or its value as a reliable pollution indicator. These roles are not mutually exclusive, although no one has been quantitatively established (21,22).

Only about twenty-five percent of this extremely diverse group of substances, some of which are hygroscopic, has, in molecular terms, been identified. The greatest source of atmospheric particulate matter (aerosols) is nature; the wind erosion of land and sea produces a world total of 3-4 million tons of aerosols per day. Another million tons of aerosols arise from vegetation and a similar quantity is derived each day from man-made sources, e.g. combustion of fossil fuels, industrial processes, and gas-phase reactions (23). The man-made aerosols, while constituting only about 20 percent of the total, are of greater concern, if for no other reason than their prominence in populated areas.

Among the metallic elements which have been identified in urban air are iron, lead, zinc, cadmium, copper, aluminum, arsenic, potassium, sodium, vanadium, molybdenum, and manganese. These tend to occur with maximal air concentrations in excess of 0.1 $\mu g/m^3$. Much of the inorganic phase of suspended matter consists of siliceous minerals (e.g. quartz, asbestos) and of soot.

Hundreds of aliphatic and aromatic hydrocarbons have been identified in the organic phase of urban aerosols and most of these are due to incomplete combustion of fossil fuels. One class of organic substances which has received considerable attention is that of the polycyclic hydrocarbons, e.g. pyrenes, naphthenes, etc., because of their carcinogenic actions. The carcinogen, 3,4 - benzapyrene, has a peak urban concentration of about 0.1 $\mu g/m^3$ (24,25).

GASEOUS POLLUTANTS -- The common gaseous pollutants in urban air are the low molecular weight hydrocarbons, and the oxides of sulfur, carbon, and nitrogen, all of which are chiefly derived from the combustion of fossil fuels. The photochemical production of ozone by sunlight and its subsequent reactions with various pollutants, produce nitrogen dioxide (NO_2) and many "activated" forms of the hydrocarbons. These substances, i.e. ozone and NO_2, are members of a special group called the oxidants, because they are vested with an oxidative capability.

Sulfur dioxide (SO_2) and trioxide (SO_3) are the major air pollution forms of sulfur, but various sulfides and mercaptans (sulfur alcohols) are always detectable.

Nitrogen oxide (NO) is produced by all high temperature, combustion processes involving air, e.g. that within the internal combustion engine. Nitric oxide is readily oxidized to nitrogen dioxide by a photochemical reaction (see oxidants). Since the persistence of NO is rather brief, NO_2 is generally considered to be the important oxide of nitrogen in air pollution. Although not as strong an oxidant as ozone, the toxicology of NO_2 is qualitatively similar to ozone (26,27).

Carbon monoxide (CO) can be viewed as an incomplete oxidation product of hydrocarbons and other carbonaceous materials. Carbon monoxide is an asphxiant, which implies that it interferes with oxygen utilization in the organism. Its companion gas, carbon dioxide (CO_2), is the ultimate combustion product but, while produced in greater tonnage than the other gaseous pollutants, it is relatively harmless to man. The carbon dioxide cycle in the environment has been qualitatively established, but the ultimate impact of carbon dioxide directly or indirectly on man depends upon the capacity of several components in the cycle, especially that of the oceans.

In summary, we can conclude that the urban air pollutants which have a well associated relationship with respiratory dysfunction of disease are: (a) air pollution generally, including various kinds of smogs; (b) oxidant gases and, in particular, ozone at the highest levels which have been identified in urban areas (28); and (c) carbon monoxide, on the basis that at the highest urban concentrations there is sufficient hypoxia to stress cardiopulmonary patients (29). In all of the foregoing situations, the susceptible individuals in the urban environment are the subjects at risk.

OCCUPATIONAL POLLUTION

The study of occupational pollution is, by its nature, varied, but in general, both epidemiology and animal experimentation have played major roles in determining the related respiratory effects. Since most industrial exposures involve only employees, who constitute a select population, there is a need to screen all process materials by animal studies, and to institute controls against inhalation exposures according to the toxicological classifications revealed. On the other hand, where a material is utilized in many kinds of industries, and often discharged in industrial effluents, then employee histories and population studies are usually required.

Since industries are among the major energy consumers, they have, of course, a requirement for large amounts of fossil fuels and consequently produce enormous amounts of the pollutants found in urban air. They tend, moreover, to utilize fuels with a higher sulfur content than that used by the automobile. Consequently, industrial areas tend to have disproportionately higher sulfur oxide levels (30). But perhaps more significant is the industrial generation of air contaminants of a special toxicological type. An outstanding example is asbestos, which, in the United States, is mainly chrysotile. Although small amounts of asbestos in urban areas have been traced to automobile brake linings, the most important sources are asbestos production and use. In the latter category, the use of asbestos in building construction is a good example. Asbestos exposures tend to be somewhat peculiar in affecting not only the individuals intimately concerned with its production and use, but also those having rather casual contact with it (31). Characteristically, there may be a health problem in a reasonably well demarcated district, although the history of exposure is often bizarre and indistinct, and reminiscent of the lung problems associated with beryllium oxide production (32).

Other air contaminants commonly having their origins in industrial or trade operations are cement, iron oxide, ammonia compounds, alcohols, aldehydes, ketones and acids, fluorine and fluorides, natural and synthetic fibers, plastic monomers, oil mists, coal dust and fly ash, and metal fumes and vapors (e.g. mercury and lead) (33,34).

A variety of occupational lung diseases are associated with dust, fumes, and other noxious materials inhaled by industrial workers and miners (33). They are typified by effects ranging from irritation of the respiratory membranes to severe pathological changes such as fibrosis and edema, all of which can be complicated by pneumonias and other serious respiratory infections. The diseases produced by mineral dust, solvent vapors, and irritant gases are among the most common; in addition, a type of pulmonary disease is produced by chemical sensitization, i.e. through an immunological mechanism. Example of such agents are toluene diisocyanate, a foaming agent used in the plastics industry, many organic dusts such as animal hair, and some inorganic substances such as beryllium oxide (Chapter 13). It is always possible that some individuals living in the immediate environs of industries or mines may show the milder forms of various diseases.

Among the well established diseases and agents are stannosis, siderosis, and baritosis which result from the processing of tin oxide, iron oxide, and barium oxide, respectively. These pneumoconioses are essentially X-ray diagnosed in that functional impairment is negligible. Asbestosis, berylliosis, talcosis, and silicosis caused by asbestos, beryllium, talc (magnesium silicate), and crystalline silica exposures, respectively, are typical of the more serious fibrosing diseases. Both coal and kaolin (a silicate) workers are subject to respiratory diseases which vary widely in their severity, possibly due, in part, to the presence of other minerals, such as silica. The production of manganese, cadmium, zinc, vanadium, and aluminum are also associated with important respiratory dysfunctions and disease, generally ranging from a chronic bronchitic phase to progressive dyspnea and fibrosis (33,35,36).

There are also respiratory diseases caused by the use of natural products in certain trades as, for example, the following conditions: bagassosis due to moldy bagasse; farmer's or mushroom worker's lung due to thermophilic actinomycetes (e.g. Micropolyspora sp.); byssinosis and weaver's cough caused by cotton fibers and allied dust; maple bark stripper's disease, due to the fungus, Cryptostroma corticale; and various forms of extrinsic allergic alveolitis which have been traced to allergenic, organic materials (36,37).

Mining operations involve many industrial materials, or their antecedents, along with important amounts of silica (crystalline SiO_2) and silicates in their effluents. In most mines, the naturally-occurring uranium and thorium compounds give rise to a radioactive gas, radon, which decays through a series of radioactive progeny. Stacked rubble above grounds taken from such mines, or building materials made from it, may also give rise to high ground level concentrations of radon and its decay products. The radon progeny are highly suspect of causing lung cancer in uranium miners (38), and should be monitored in other situations where people are subject to exposure.

The expanding use of nuclear energy has been accompanied by an increased handling and use of radioactive substances (e.g. uranium, plutonium, and thorium) and, consequently, an increased occurrence of radioactive tritium, iodine, and various fission products, in reactor effluents and in the effluents of atomic fuel processing plants. All radioisotopes which tend to persist in the lungs, such as the actinide elements, are capable of producing serious lung injury, which qualitatively resembles the injury from external radiation, ranging from a pneumonitis to a diffuse progressive fibrosis. At lower exposure levels, particularly those sustained over a long time, there is the attendant possibility of increased malignancies of the lungs, but this is highly speculative.

The control of reactor environments is ordinarily designed to keep the concentration of airborne radioactivity, at a point where there is "loss of control", within the maximal permissible concentration (MPC air) of the specific emitters for the general population. "Loss of control" is generally considered to be at the geographical limits of the reactor facility and accordingly, if control is adequate, a population in the vicinity of a reactor could never be exposed to more than MPC levels. Unfortunately, validation of MPC levels in the general population is lacking.

Agricultural operations, which now center about large farms and often utilize wide-spread applications of fertilizers and economic poisons, e.g. pesticides, herbicides, and insecticides, suggest the likelihood of airborne contamination which may affect both farm workers and individuals in surrounding areas. The extent of this problem is not easily assessed since delivery is often variable (sometimes by airplanes) and the types of materials dispersed are usually complex, with hundreds of compounding variations available commercially. In some farms where silage is produced and stored in large amounts, silage gases, e.g. carbon monoxide and nitrogen oxides, are often produced in lethal concentrations although generally well confined.

In summary, it is evident that many industrial contaminants are identical to those produced by cars and stationary sources in urban areas. Asbestos appears to be a major problem; it is a carcinogenic material (producing mesothelioma and other tumors) capable of affecting populations outside of the industry (31). In limited areas, odor and fume problems, e.g. irritation of the respiratory membranes, are well known. During major atmospheric inversions many toxic chemicals may build up in industrial areas and add to the existing problem. This is largely presumptive, but studies are underway on the impact of industrial pollution on community respiratory illnesses (Chapters 13 and 15).

NATURAL POLLUTION

As pointed out earlier, on a global scale, the major sources of atmospheric aerosols are natural; specifically, the wind erosion of land and sea masses produces millions of tons of largely inorganic aerosols each day. Prominent in this group of aerosols are sea salts, e.g. sodium chloride, and siliceous materials, especially silicates, derived from soil. Also, vegetation contributes many kinds of organic dusts and gases by direct injection, e.g. pollen, spores, terpenes, or by decay and comminution, e.g. fungi, methane, ammonia, hydrogen sulfide, and organic debris.

From the viewpoints of allergic and infectious respiratory disease, the nature, abundance, and atmospheric transport of airborne microbes are of considerable interest and were the subject of a recent symposium (39).

Many of the volatile components appear to enter into gas-phase reactions (noted earlier for the common urban air pollutants), with the result that compounds such as ammonium nitrate and ammonium sulfate are spontaneously formed as extremely small particles called nuclei. Also, to the extent of tons per square mile per month, particulate matter of extraterrestrial origin settles on the earth surface.

The effect of air contaminants of natural origin on the human respiratory system has not been studied extensively, probably because the airborne levels of these contaminants are ordinarily small in terms of other human exposures. Obviously, this statement is more relevant to urban populations than rural. Wide-scale meteorological effects and changes in atmospheric visibility are unquestionably relatable to natural air pollution; for example, the persistent haze seen in the Smoky Mountains is due to terpenes from conifers, but in the health effects area, only naturally-occurring, allergenic substances such as pollens are important and causally associated with respiratory distress, e.g. hay fever, allergic asthma. This information has been derived from immunological testing and by epidemiology wherein atmosphere pollen indices have been correlated with morbidity (37) (Chapter 15).

PERSONAL POLLUTION

Perhaps the foremost form of air pollution encountered by modern man is that associated with tobacco smoking. A simple description of tobacco smoke is that it is an aerosol containing carbonaceous particles, many kinds of organic substances and tars, and relatively high concentrations of certain gases such as nitrogen oxides and carbon monoxide. Tobacco smoke, both chemically and physically, varies according to the additives present (sugar, aromatics), the burning temperature, the age and concentration of the smoke, the type of tobacco leaf used, the presence of a filter, and a multitude of other factors. Smoking is probably the only form of personal pollution which has been intensively investigated (Chapter 12). It is highly suspect of aggravating, if not causing, many forms of respiratory diseases, viz. emphysema, lung cancer, and chronic bronchitis.

Aerosol sprays are a common source of air contaminants in the home. Most households have hair sprays, oven cleaners, touch-up paints, furniture polish, insecticides, and deodorant preparations dispersed from pressurized spray cans. The foregoing list would suggest aerosols of animal protein; caustic alkalis, or alkaline detergents; alkyd resins; lemon oil and vegetable waxes; formaldehyde and aluminum chlorhydrate; and pyrethrumns, respectively, all associated with an assortment of petroleum distillates, perfumes, and vehicles (34). Collectively, these diverse preparations are probably all respiratory irritants, but little or no toxicological information exists, nor are most aerosol dispensers labelled as to their contents. Fortunately many aerosol sprays are so coarse as to be largely non-respirable.

Air contamination is also associated with many kinds of manufactured powders, particularly when the particle size is small and the powder is dry. This characterizes many of the modern detergents (40,41). Volatile dry cleaning preparations and several kinds of household cements ("airplane glue") (42) usually containing chlorinated hydrocarbons, are also potential household contaminants.

In summary, many potential and established respiratory hazards exist in the home. Occasional cases of asthma, bronchitis, pulmonary edema, and other types of sensitivity reactions have accompanied the use of enzyme detergents and aerosol sprays. The size of this problem is unknown, but it is probably of sufficient magnitude to warrant investigation and control. Tobacco smoking, in any case, looms as one of the most destructive forms of air contamination and in any context, for example, the study of urban pollution on the health of miners, the smoking habits of the populations concerned may overwhelm the affects being assessed (Chapter 12).

ENTRY INTO AND REMOVAL FROM THE RESPIRATORY SYSTEM

In succeeding sections of this report, many of the interactions between airborne contaminants and the functional anatomy of the human respiratory system will be described. To lay a basis for these interactions, we should first examine the initial encounter of airborne contaminants with the respiratory system. In this regard, we will see that the physical and chemical properties of the airborne contaminants, the peculiar features of cyclic breathing, and our complex respiratory anatomy work together to produce differences in respiratory exposure, both qualitatively and quantitatively. Two quantitative measures of dosage, input and intake, must be distinguished. Input is an estimate of the amount that enters the respiratory tract, based on such data as the concentration of the agent in the air that is breathed in and the volume of air respired per minute. Intake is the amount of material actually taken up from the inspired air by the individual, or, more cogently, the amount that arrives at the susceptible tissue site. The intake varies with such factors as the intimate distribution of the air-borne agent in the respiratory tract, its deposition on the walls, its ability to pass through the walls to reach the target tissue, etc. It is clear that the ratio of intake to input can vary greatly from one individual to another, and in the same individual from one time to another. In general, for airborne gases and particulates, the intakes are almost always less than the inputs.

With aerosols, the inherent instability of a particle suspension in air brings about deposition. If the particles are relatively large and dense, they experience gravitational settling. If they are placed in an air stream and suddenly forced to make sharp turns, their inertial properties are sufficient to cause the particles to cross the streamlines, that is, they cannot follow the air stream. When this condition occurs in a tubular system, e.g. the airways of the respiratory tract, the particles inertially impact on the tube walls. Particles which are small enough to be airborne, yet large enough to exhibit the foregoing behavior, have sizes between ca. 1 and 200 μm diameter (at least for particles of density 1 g/cm^3 or thereabouts).

When particles are below microscopic resolution (equal to or less than 0.4 μm), they have negligible gravitational and inertial properties. Particles below 500 Ångstroms (0.05 μm) are greatly affected by the random thermal motion of air molecules and consequently diffuse and deposit on surfaces; this is often termed Brownian motion deposition. As the particle diameter decreases further and approaches that of gas molecules, the particle's diffusivity increases and its ability to remain airborne decreases.

When a subject breathes an aerosol cloud, he usually brings in particles of widely different particle size and, accordingly, of suspensions with widely differing stabilities. The more unstable the suspension, i.e. the greater the inertial, gravitational, or Brownian motion tendencies, the greater the probability that a particle will deposit within the respiratory system. For the larger particles, deposition occurs principally in the upper airways and the tracheobronchial tree. These are the areas of relatively high air velocity with many changes in the direction of air flow. For particles above 10 μm diameter, deposition is virtually absolute, with the nasal passages being a dominant deposition site. Submicronic particles easily penetrate into all parts of the respiratory system and they continually experience Brownian motion deposition, which quantitatively predominates in the alveolar region. A certain proportion of submicronic particles remain airborne and are expelled by expiration. A generalizing schematic of particle-deposition probabilities as a function of particles size and respiratory regions is depicted in Figure 1.

It should be noted that all particulate collisions with surfaces are regarded as inelastic and that particles once deposited cannot be easily resuspended. Resuspension can occur during coughing

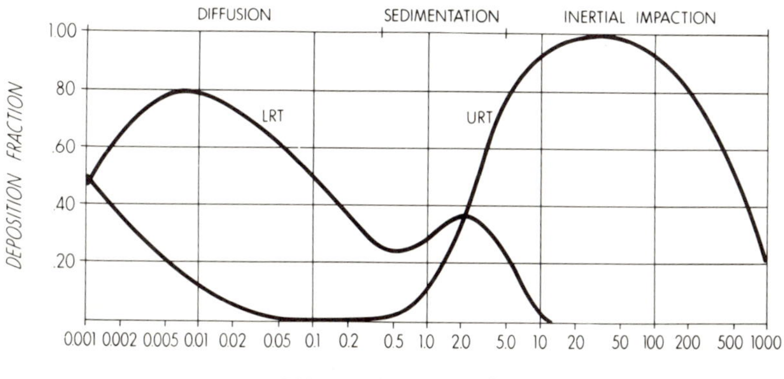

Fig. 1. Particle Size Deposition Probabilities in the Human Respiratory System. The URT curve pertains to deposition in the nasal passage and other supraglottic structures. LRT deposition is that which occurs in the tracheobronchial tree and lung parenchyma. The proportion of particles of a particular size which remain undeposited, and will be expired, is obtained by subtracting the sum of the values on the two deposition curves from 100.

and sneezing whereupon particulate-laden secretions are made airborne. This is normally a minor and irregular particulate removal mechanism.

Gas deposition differs from particulate deposition in that the "instability" of gas molecules is due to their diffusivity and their reactivity; there are no gravitational and inertial effects of importance and most collisions are elastic (reversible). In the respiratory system, a nonreactive gas differs in its penetration and mixing, when compared to air, only by virtue of its different molecular weight or viscosity. With reactive gases, such as, a gas which is highly soluble in water, e.g. sulfur dioxide, the normally elastic encounters which would be characteristic of dry-surface collisions is markedly changed when interacting with a wet surface, such as the respiratory membrane. The large surface area of wet mucous membrane in the upper airways, and the high aqueous solubility of SO_2, lead to an almost complete absorption of the gas before it can penetrate into the lungs. Ozone, an oxidant gas, reacts chemically with a host of materials and, although its water solubility is relatively limited, the end result is qualitatively similar to that of SO_2, in that a high inspiratory concentration of ozone is greatly reduced by the time it reaches the alveolar level by virtue of its oxidative reactions with the respiratory surfaces along the way.

Once deposited, gases may be chemically "fixed", reversibly or irreversibly, or they may be distributed about the lung and body as a dissolved phase; otherwise, generalizing features of gas clearance are difficult to describe.

Particulate clearance from the lungs and airways appears to be dependent onthe material involved and the region of deposition. The upper and lower airways are chiefly cleared by the motion of the mucous blanket imparted by microscopic hairs called cilia. The general drift of this blanket is toward the gastrointestinal tract. If the particle is "soluble", it may be absorbed from the airways which are richly vascularized, or it may dissolve in the mucous secretions and move as an insoluble particle would.

Particulate clearance from the lung parenchyma, which is non-ciliated, appears to involve an absorptive mechanism whereby the particle or its dissolved phase moves into the blood or lymph. This mechanism appears to depend on permeability considerations and endocytosis; the latter involves the transport of particles by cellular actions. One type of endocytosis results in the transfer of particles by mobile cells called macrophages. These migratory cells engulf particles and transport them out of the lungs by several pathways (Chapters 2,4). Details of alveolar clearance are lacking and especially quantitative data. Notwithstanding, alveolar clearance of insoluble particles, unlike their clearance from ciliated airways, which involves minutes or hours, may vary from hours to years, depending upon the particulate material. As a result, small amounts of some materials may have progressive or cumulative effects on the respiratory structures.

Studies of aerosol deposition and clearance in man, expounding upon these briefly described features, are available (43-46).

REFERENCES

1. Firket, J. (1960). Sur les causes des accidents survenus dans la vallée de la Meuse lors des Brouillards de Decembre, 1960. Bull. Acad. Roy. Med. Belg. 11: 683.

2. Schrenk, H. H. et al. (1949). Air Pollution in Donora, Pennsylvania: epidemiology of the unusual smog episode of October 1948. Pub. Health Bull. No. 306, Washington.

3. Public Health Medical Service. (1954). Morbidity and mortality during the London fog of December 1952. Report No. 95, Her Majesty's Stationery Office, London.

4. Greenburg, L. et al. (1967). Air pollution, influenza and mortality in New York City. Arch. Env. Health 15: 430.

5.   Stocks, P. (1952). Epidemiology of cancer of the lung in England and Wales. Brit. J. Cancer 6: 99.

6.   Dohan, F. C. (1961). Air pollutants and incidence of respiratory disease. Arch. Env. Health 3: 387.

7.   Ishikawa, S., Fisher, V. and Wyatt, J. P. (1969). The "emphysema profile" in two midwestern cities in North America. Arch. Env. Health 18: 660.

8.   Lunn, J. E. (1965). Respiratory measurements of 3,556 Sheffield school children. Brit. J. Prev. Soc. Med. 19: 115.

9.   Fairbairn, A. S. and Reid, D. D. (1958). Air pollution and other local factors in respiratory disease. Brit. J. Prev. Soc. Med. 12: 94.

10. Winkelstein, W. (1966). Effects of social class on respiratory morbidity and mortality in Buffalo. AMA Air Pollution Medical Research Conference, Los Angeles.

11. Verma, M. P., Schilling, F. J. and Becker, W. H. (1969). Epidemiological study of illness absence in relation to air pollution. Arch. Env. Health 18: 536.

12. Remmers, J. E. and Balghum, O. J. (1966). Effects of Los Angeles urban pollution upon respiratory function of emphysematous patients. AMA Air Pollution Medical Research Conference, Los Angeles.

13. Motley, H. L., Smart, R. H. and Leftwich, C. I. (1959). Effect of polluted Los Angeles air (smog) on lung volume measurements. J. Amer. Med. Assoc. 71: 1469.

14. Young, W. A., Shaw, D. B. and Bates, D. V. (1964). Effects of low concentrations of ozone on pulmonary function in man. J. Appl. Physiol. 19: 765.

15. Gardner, M. B. (1966). Biological effects of urban air pollution: III. Lung tumors in mice. Arch. Env. Health 12: 305.

16. Coffin, D. L. (1970). Study of the mechanisms of the alteration of susceptibility to infection conferred by oxidant air pollutants. In: M. G. Hanna, P. Nettesheim and J. R. Gilbert, (eds.). Inhalation Carcinogenesis. U. S. Atomic Energy Commission, Division of Technical Information Extension, Oak Ridge, Tennessee, 259.

17. Erlich, R., Henry, M. C. and Fenters, J. (1970). Influence of nitrogen dioxide on resistance to respiratory infections. In: M. G. Hanna, P. Nettesheim and J. R. Gilbert, (eds.). Inhalation Carcinogenesis. U. S. Atomic Energy Commission, Division of Technical Information Extension, Oak Ridge, Tennessee, 243.

18. Amdur, M. O. (1959). The physiological responses of guinea pigs to atmospheric pollutants. Int. J. Air Poll. 1: 170.

19. Hurst, D. J., Gardner, D. E. and Coffin, D. L. (1970). Effect of ozone on acid hydrolases of the pulmonary alveolar macrophage. J. Reticuloendothel. Soc. 8: 288.

20. Mendenhall, R. M. and Stokinger, H. E. (1962). Films from lung washings as a mechanism model for lung injury by ozone. J. Appl. Physiol. 17: 28.

21. Amdur, M. O. (1969). Toxicologic appraisal of particulate matter, oxides of sulfur, and sulfuric acid. Air Poll. Control Assoc. J. 19: 638.

22. Anderson, D. O. (1967). Effects of air contamination on health. Canad. Med. Assoc. J. 97: 528.

23. Hidy, G. M. (In press). The dynamics of aerosols in the lower troposphere. In: T. Mercer, W. Stöber and P. Morrow, (eds.).

Assessment of Airborne Particles, Proc. 3rd Rochester Int. Conf. of Environmental Toxicity. Charles C. Thomas, Springfield, Illinois.

24. Lindsey, A. J. (1962). Some observations on the relative quantities of air pollutants in various locations. Symposium on Analysis of Carcinogenic Air Pollutants, National Cancer Institute Monograph No. 9, 235.

25. Sawicki, E. et al (1962). Polynuclear aromatic hydrocarbon composition of the atmosphere in some large American cities. Am. Ind. Hyg. Assoc. J. 23: 137.

26. National Air Pollution Control Administration. (1969). Air Quality Criteria for Sulfur Oxides. U. S. Department of Health, Education, and Welfare, Washington, D. C. NAPCA Publ. AP-50.

27. National Air Pollution Control Administration. (1970). Air Quality Criteria for Photochemical Oxidants. U. S. Department of Health, Education, and Welfare, Washington, D. C. NAPCA Publ. AP-63.

28. Tabershaw, I. R., Ottoboni, F. and Cooper, W. C. (1968). Oxidants: Air quality criteria based on health effects. J. Occup. Med. 10: 464.

29. National Academy of Sciences. (1969). Effects of Chronic Exposure to Low Levels of Carbon Monoxide on Human Health, Behavior, and Performance. Washington, D. C.

30. National Air Pollution Control Administration. (1970). Nationwide Inventory of Air Pollutant Emissions, 1968. U. S. Department of Health, Education, and Welfare. NAPCA Publ. AP-73.

31. Tabershaw, I. R. (1968). Asbestos as an environmental hazard. J. Occup. Med. 10: 32.

32. Mancuso, P. F. (1965). Biological effects of asbestos. Ann. N. Y. Acad. Sci. 132: 589.

33. Patty, F. A. (ed.). (1958). Industrial Hygiene and Toxicology, Vols. I, II, and III. Interscience Publishers, New York.

34. Gleason, M., Gosselin, R. and Hodge, H. C., (eds.) (1963). Clinical Toxicology of Commercial Products. Williams and Wilkins (2nd ed.), Baltimore.

35. Worth, G. and Schiller, E. (1954). Die Pneumokoniosen. Staufen-Verlag, Köln, 573.

36. Crofton, J. and Douglas, A., (eds.). (1969). Respiratory Diseases. Blackwell, Oxford.

37. Parish, W. E. and Pepip, J. (1963). Allergic reactions in the lung. In: P. G. H. Gall and R. R. A. Coombs, (eds.). Clinical Aspects of Immunology. Blackwell, Oxford.

38.* International Atomic Energy Agency. (1969). Radiation-Induced Cancer. STI/PUB 228, Vienna.

39. Gregory, P. H. and Monteith, J. L., (eds.). (1967). Airborne Microbes, Seventeenth Symposium of the Society for General Microbiology. Cambridge Univ. Press, New York.

40. Fulwiler, R. D. (1971). Detergent enzymes -- An industrial health challenge, Am. Ind. Hyg. Assoc. J. 32: 73.

41. Newhouse, M. L. et al. (1970). An epidemiological study of workers producing enzyme washing powder. Lancet I: 689.

42. Reinhardt, C. F. et al. (1971). Cardiac arrhythmias and aerosol sniffing. Arch. Env. Health 22: 265.

---

* See also Lundin, F. E., Wagoner, J. K. and Archer, V. E. (1971). Radon Daughter Exposure and Respiratory Cancer Quantitative and Temporal Aspects. NIOSH-NIEHS Joint Monograph No. 1. U. S. Public Health Service, Washington. Publised since References were compiled (Ed.).

43.   Morrow, P. E. (1970). Models for the study of
particle retention and elimination in the lung.
In: M.   G.   Hanna, P. Nettesheim and J.   R.
Gilbert, (eds.). Inhalation Carcinogenesis.   U.
S.   Atomic Energy Commission, Division   of
Technical   Information Extension, Oak Ridge,
Tennessee, 103.

44.   Lippmann, M. and Albert, R. E.   (1969). The
effect   of   particle   size on   the regional
deposition of inhaled aerosols in the human
respiratory tract. Am. Ind. Hyg. Assoc. J.
30: 257.

45.   Bates, D. V. et al. (1966).   Deposition and
retention models for internal dosimetry of the
human respiratory tract.   Health Physics 12:
173.

46.   Morrow, P. E. (1971). Alveolar clearance of
aerosols. Arch. Int. Med. (In press).

CHAPTER 7. ACTION THROUGH CHANNELS OTHER THAN AIRBORNE

ARON B. FISHER, Departments of Medicine and Physiology,
University of Pennsylvania School of Medicine,
Philadelphia, Pennsylvania

ROBERT FRANK, Department of Environmental Health,
University of Washington School of Public Health
and Community Medicine, Seattle, Washington

H. KENNETH FISHER, Department of Medicine,
Harborview Medical Center, University of Washington,
Seattle, Washington

The number of environmental pollutants that may reach and affect the lung through channels other than inspired air is small, but many substances that are self-administered or given for therapeutic reasons can affect the lung, or its reactions to air-borne contaminants. The main modes of entry that may be distinguished are: through the skin directly or by puncture, through the alimentary canal, or by aspiration into the airways from the mouth or pharynx. These routes of transport will be examined, and then the principal effects on lung function will be discussed. Special mention will be made of the interesting effects of the herbicide, paraquat.

AGENTS ENTERING THROUGH THE SKIN

While some substances are absorbed through the skin, injection is likely to result in more rapid transport to distant organs such as the lung. If the injection is directly into a vein or lymphatic, the transport will be particularly rapid. Injection into an artery, however, may result in absorption or modification in the peripheral capillaries, and delay of arrival at the lung.

TRANSPORT BY VENOUS SYSTEM

Injection into a peripheral vein is the most direct means other than airborne by which material can be transported to the lungs. Because the pulmonary capillaries will be the first vascular bed exposed to the agent, a very high concentration of the substance will be delivered to the pulmonary capillary endothelium. Since the toxic action of a substance on the lung might depend on a dose-duration relationship, some of the factors that determine the dose of agent reaching various parts of the lung and the duration of contact between agent and lung tissue will be examined.

The fraction of the injected dose that reaches any part of the pulmonary capillary bed will depend on the intrapulmonary distribution of the blood containing the substance. This distribution will be primarily determined by the pulmonary artery pressure together with the local vascular resistance. As an example of the influence of these physiological parameters, if the injection is made with the subject upright, most of the foreign material will be delivered to the lung bases, whereas if the subject is supine, the distribution will be more uniform. The dose-duration relationship of the injected material will also be influenced by the fate of the toxic material in the lung circulation. Large particles such as cotton fibers or talc might be trapped in the arterioles and cause infarction or granulomatous arteritis (1,2). Volatile substances such as ether or carbon tetrachloride pass across the alveolar-capillary septum and are excreted in the expired air, but may damage the cells comprising the septum. Some drugs, such as the vasoactive agents, histamine, serotonin, bradykinin and prostaglandins, appear to be partly removed from the circulation by the lung cells and metabolized (3). The majority of agents, though, are probably passively distributed in the pericapillary space of the alveolar septum during passage through the capillaries. The volume of lung substance through which the foreign agent is distributed depends on the vascular permeability to that molecule and its solubility in the various tissues comprising the lung parenchyma. The volumes of distribution for substances of various size and solubility have been studied by the multiple indicator technique (4). However, this information is not available for most of the potentially toxic agents. In addition, capillary damage by a toxic substance may change the vascular permeability and thereby alter its volume of distribution.

Several toxic reactions in the lung probably depend on the effects of these peak dose-duration relationships on the pulmonary vasculature. For example, experimental production of pulmonary edema by the injection of alloxan is probably due to direct capillary toxicity related to dose (5). The response to the vasoactive amines, e.g. histamine and serotonin, is also dose related (6). Clinically, the pulmonary edema that may develop with heroin overdosage is probably on this same basis (7), although alternate explanations of a capillary hypersensitivity reaction, pulmonary venous spasm or central nervous system mediated response cannot be excluded.

After the intravenously injected bolus has passed through the lung recirculation occurs, but by this time the toxin has been diluted in the systemic circulation. With subsequent passages, the concentration of the substance will approach a value determined by its volume of distribution in the whole body and its rate of removal by excretion or metabolism. In the steady-state situation, the concentration reaching the lung should be similar to that which occurs after absorption from the gastrointestinal tract.

In addition to reaching the lung via the pulmonary capillaries, the drug may be transported by the other vascular supply to the lungs, the bronchial arteries. Through this circulation potentially toxic substances can reach the airways down to the level of the respiratory bronchioles. However, the blood flow and, therefore, the amount of material carried by the bronchial vessels in normal persons, is only about 1% of what passes through the pulmonary capillaries, although higher values occur with disease (8). This may be the route by which bronchospasm due to hypersensitivity results from systemic administration of a drug.

TRANSPORT BY LYMPHATIC SYSTEM

Material infused into a lymphatic vessel will enter the blood stream via the thoracic duct after passing through the lymph nodes of the involved chain. Again the lungs represent the first visceral capillary bed exposed to the drug. Diagnostic lymphangiography with iodized oils is the only common use of lymphatic injection. This procedure has been shown to result in decreased pulmonary diffusing capacity and lung compliance, probably due to mechanical blockage of the capillaries by oil droplets (9).

AGENTS ENTERING FROM THE ALIMENTARY CANAL

BY ABSORPTION FROM THE GASTRO-INTESTINAL TRACT

This is the most common route by which toxic substances enter the blood stream. In contrast to intravenous injection, the foreign substance is usually exposed to two capillary beds, gastrointestinal and hepatic, before reaching the pulmonary capillaries. Therefore, direct toxic effects on the pulmonary capillaries are unlikely to be the major clinical manifestation of poisoning. An exception occurs in the case of the weed-killer paraquat, which induces interstitial and alveolar edema in the lung followed by proliferation of fibroblasts and other cell types (10). However, most lung effects from alimentary absorption are due to hypersensitivity or idiosyncratic reactions. The pathophysiologic manifestations include acute bronchospasm due to bronchial allergy to food substances or drugs; pulmonary and pleural fibrosis from chronic administration of the drug methysergide (11); acute pulmonary edema, probably due to hypersensitivity reaction of pulmonary capillaries, with nitrofurantoin and salicylates (12,13); chronic alveolar edema leading to alveolar fibrosis with busulfan, hexamethonium, and mecamylamine (14,15); and pulmonary inflammatory changes, frequently with peripheral eosinophilia (Loeffler's syndrome), after penicillin, sulfonamides, para-amino salicylic acid, mephenesin, and methotrexate (16-19). The lungs can also be involved as part of a diffuse systemic vasculitis or systemic lupus erythematosus induced by hydantoin derivatives, hydralazine, and procainamide (20).

In contrast to these direct cause and effect reactions are several examples of a less direct relationship. For example, there is a high incidence of pneumonia with agents that depress pulmonary defense mechanisms. Instances are the depression of alveolar macrophage phagocytic ability by alcohol, interference with antibody formation due to antimetabolites, and alteration of cellular immunity by adrenal cortical steroids (21,22). A second type of indirect relationship exists in the suspected high incidence of thrombophlebitis, leading to pulmonary embolism, in persons taking progestational agents for birth control or other reasons (23).

BY ASPIRATION INTO THE TRACHEA

Aspiration of foreign material from the pharynx into the trachea is another route other than airborne by which toxic agents can reach the lung. In normal circumstances, reflex closure of the larynx prevents aspiration of material and the cough reflex expels material inadvertently aspirated. However, in the presence of neurological disease, debility, drug-induced depression of the central nervous system, or esophageal malfunction, the action of these protective reflexes might be impaired. Aspiration is facilitated by the supine position since material will tend to collect in the hypopharynx. The ease of aspiration is influenced by the nature of substance, e.g. a bland substance such as mineral oil is relatively ineffective in eliciting the cough reflex. Low density materials such as petroleum distillates tend to be aspirated more readily.

The distribution of the aspirated foreign material through the tracheobronchial tree will be partially dependent on gravity and, therefore, on body position. If aspiration occurs with the subject upright, the aspirate will be preferentially distributed to lower lobes. While in a supine subject the posterior segment of upper lobes and superior segment of lower lobes may be predominantly affected. The clinical example of lipoid pneumonia due to mineral oil illustrates the effect of position, since the material is usually aspirated when the person is supine and the lesions tend to occur in the mid-lung field (24).

The reaction of the lung depends on the material aspirated. Large foreign bodies may cause obstructive atelectasis. Kerosene, turpentine, furniture polish, and lighter fluid, which are frequent causes of childhood poisoning, may be easily aspirated because they induce vomiting and are low density substances -- aspiration of one of these substances causes severe pulmonary inflammatory change which is the usual cause of death in fatal poisoning (25,26). Mineral oil, another petroleum product frequently aspirated, is much less toxic to the tissues; aspiration of this material causes subacute inflammatory changes and fibrosis (27). This latter syndrome of "chronic lipoid pneumonia" is seen most frequently in elderly persons who use oily nose drop preparations or swallow mineral oil at bedtime.

DIRECT TRANSMISSION THROUGH THE CHEST WALL

The final means by which environmental agents can reach the lung is by direct transmission through the tissues of the chest wall. High energy irradiation from the short end of the electromagnetic spectrum, i.e. gamma and x-rays, can reach the lungs by this route. With whole-body irradiation from an atomic blast or accidental spillage, lung damage is unlikely to be of primary concern since other tissues are more susceptible to radiation damage (28). However, therapeutic irradiation of tumors of the breast, lung, or mediastinum may result in the syndrome of radiation pneumonitis (29), although current techniques of supervoltage therapy have reduced the danger. The postulated mechanism for radiation damage is the formation of free radicals by the ionizing radiation, with toxic changes in the cells comprising the alveolar septum. The endothelial cells appear to be most susceptible to these effects, but necrosis and sloughing of Type 1 epithelial cells may also occur (30). Subsequently, collagen proliferates in the damaged area and results in pulmonary fibrosis.

Barometric pressure waves can also be transmitted through the chest wall to the lungs. Chest wall transmission is not important during slow changes in barometric pressure as occurs with diving or ascent (pulmonary effects in these circumstances are airborne). However, with rapid change in ambient pressure as occurs during air blast or underwater detonation, transmission of the pressure wave through the chest wall may be responsible for pulmonary hemorrhage and tearing of alveoli (31). Available evidence indicates that the damage occurs as the pressure wave travels from the water-density tissues of the thorax to the air-density lung tissue (32). The peak pressure and the pressure integral both seem important in determining the extent of damage (33).

SECONDARY EFFECTS FROM ENVIRONMENTALLY INDUCED LESIONS ELSEWHERE

Because of the central position of the lungs with respect to the blood circulation, disease of virtually any organ may secondarily affect the lungs. Thus, pulmonary edema or pleural effusions may result from cardiac or renal failure induced by drugs; lung metastases may occur from cancer induced elsewhere; shock might have secondary effects on the lung parenchyma; central nervous system lesions may induce abnormalities of both pulmonary function and structure (34). In this context, then, the lung might be secondarily affected by a great many environmental agents that do not have direct toxic action on pulmonary tissue.

ACTION ON THE LUNG

Damage to the lung by environmental agents that enter the body through the gut or skin is probably infrequent, but can be serious. Like the kidney, the lung is an organ of excretion. If a volatile organic compound such as carbon tetrachloride (CCl_4) is ingested, a fraction of the amount reaching the circulation is then excreted as a vapor. The excretion of the vapor may be accompanied by alveolar injury. Like the liver, but on a more modest scale, the lung is also a complex metabolic system. Its metabolic functions may be stimulated or depressed by agents transported in the circulation. If the activity of a specific detoxifying enzyme is impaired, the lung may as a consequence be rendered more vulnerable to inhaled carcinogens. Some examples will be given of each of the two modes of damage: the one related principally to excretion, the other consequent upon some intervention in metabolic function.

ACTION BY EXCRETION

CARBON TETRACHLORIDE -- Following its ingestion, CCl_4, a halogenated hydrocarbon, is excreted in part through the lung; other volatile solvents may be disposed of in a similar way. The excretion of CCl_4 vapor is associated with injury to all components of the alveolar-capillary membrane (35,36). Ultrastructural cellular changes in the lung appear within one hour of the feeding (intubation) of rats with 0.25 ml/100 g body weight. Several hours later, piloerection and lassitude become evident. Endothelial damage and thrombosis, edema, hemorrhage, and focal atelectasis tend to coincide with the periods of peak blood level and peak excretion. The most likely basis for the widespread damage is the solvent action of the CCl_4 on the lipids present in the cellular membranes and organelles.

SULFUR DIOXIDE -- (SO_2) is also mentioned to indicate how diverse the paths leading to the lung may be. The role of SO_2 in urban and industrial air pollution is considered elsewhere in this monograph. Virtually all of the SO_2 in inspired air is removed by the upper airways during quiet breathing (37). Only a minute fraction of the inhaled gas, amounting to 1% or less, is likely to reach the alveoli, especially if breathing is by nose. There is evidence, however, that a portion of the SO_2 that is adsorbed in the upper airways enters the venous circulation in physical solution and, upon reaching the lung, is excreted as a gas (38). That is, the exposure of the alveoli to SO_2 takes place "from within". Based on experiments in which radioactive SO_2 was administered to dogs, it has been estimated that about 2% of the radioactive sulfur in the mixed venous blood may be excreted as a gas into the alveoli during exposure (38). What effect this form of exposure may have on the lung is undetermined.

ACTION ON METABOLIC FUNCTION

Alkylating Agents

Alkylating agents, which introduce an alkyl group into another compound, can sometimes attack the nucleic acid portion of the genetic material DNA. It is believed that such transformations can induce cancerous development in the affected cells. Lung cancer, among others, can be produced experimentally by this means. Substances known as nitrosomines, for example, can be converted into alkylating agents in the body, and have been used to produce lung cancer in rats and hamsters.

The question has been raised lately as to whether the liberal use of nitrates and nitrites in the preservation of some foods, might not give rise under favorable circumstances in the alimentary canal to nitrosomines, and thus be able to induce cancer in various organs after absorption. It has by no means been proven that this actually does occur, but the possibility is receiving close study.

Interference with Enzyme Induction

Kotin and Falk (39) have suggested that the irregularity with which individuals develop cancer in response to the inhalation of polycyclic hydrocarbons may reflect varying ability to detoxify the carcinogenic compounds. If that is so, then substances arriving at the lung through vascular channels, as well as those coming through inspired air, may affect the enzymes which tend to destroy carcinogens. The carcinogen, 3,4 benz-a-pyrene, for example, is destroyed by the activity of the enzyme benzpyrene hydroxylase. Formation of the enzyme can be stimulated or inhibited experimentally by the injection or ingestion of appropriate drugs. Variations of this kind may be brought about by a wide variety of substances not yet adequately investigated. Whether or not cancer develops after exposure to a carcinogen would depend upon the net effect of many such factors interacting in complex fashion. Several naturally occurring constituents of plant foods, such as the flavones, tangeretin and nobiletin, may increase the hydroxylase activity of the lung.

EFFECTS OF THE HERBICIDE PARAQUAT

In 1958 some British scientists discovered that paraquat, a chemical agent that had been used in analytical procedures for 25 years, and some related compounds, were efficient herbicides. Since it was believed that they offered little danger to man, they came into widespread use for agricultural and military purposes. They act by interfering with plant photosynthesis, the process by which plants capture and store energy from the sun.

It has subsequently become clear that paraquat is by no means entirely safe for man. Although severe poisoning is not always fatal, medical reports have described nine deaths from poisoning with this agent (10,40-46), and it is likely that others have occurred without being published. Although damage is recognized first in other parts of the body, it is pulmonary damage which usually proves fatal. After a delay of some days there is air hunger, a fall in blood oxygen, and death due apparently to nearly complete collapse of the lungs. At autopsy the lungs are virtually airless and contain abnormally large amounts of fluid.

The structure of the lungs with their large number of very small gas exchange spaces (alveoli) renders them very susceptible to collapse. Indeed, collapse is avoided mainly by the presence of a specific surfactant thought to be manufactured for that purpose by the type II alveolar cells. (See Chapter 4) In 1967 Mantkelow (47) suggested that the collapsed lungs seen with paraquat intoxication might be due to interference with surfactant.

Fisher and Clements (48) studied rats, three days after injecting a standard dose of paraquat. About one-fifth had died, and survivors were obviously ill: they moved when alarmed but otherwise stood motionless -- their rapid breathing punctuated by wide-mouthed gasps. At autopsy the lungs of poisoned rats were nearly airless. By step-wise inflation and deflation with air it was shown that lungs from poisoned rats were "stiffer": during inflation-deflation with saline solution this difference was not observed, and the "stiffness" was therefore due to surface tension effects normally held in check by surfactant.

With a sensitive technique for measuring the quantity of surfactant (49) they examined material washed from the lungs through the bronchi. In lungs from 58 control rats they recovered more than enough surfactant to line all the alveolar surfaces. In lungs from nearly all of the 22 paraquat-injected rats there was too little surfactant to detect using this technique.

To resolve the question as to whether surfactant was really lacking or simply masked, they carried out analyses of lung lipid composition. Among the ten lipid classes analysed, they found that nine actually tended to be present in greater amounts in the lungs of paraquat-poisoned rats. Only the fully saturated lecithins were present in smaller amounts -- but the average reduction of only 25% seemed too small to account for the drastic changes observed in surface-active material and in lung behavior. Interference with surface-active material thus seemed the more likely cause.

Victims of paraquat poisoning have abnormally low blood oxygen values and have sometimes been treated with oxygen therapy. Details of the biochemical effects of paraquat suggest, however, that oxygen therapy might actually be harmful. The work of Davidson and Papirmeister (50) and of Fisher et al. (51) supports the concept that paraquat interferes with energy metabolism in bacteria and in rat lung tissue, as it does in green plants.

To test whether these effects rendered the whole animal more susceptible to injury by oxygen, Fisher and Clements exposed rats to an atmosphere of moist oxygen following injection with Paraquat. The results were striking (Table 1). Whereas 186 rats had all survived the standard dose for at least 24 hours and (132 of these had survived for 48 hours) breathing room air, fatalities were now observed in 19 of the 20 rats breathing oxygen within the first 22 hours after injection. A smaller number of cats, rabbits, and guinea pigs also showed remarkable pulmonary edema after 24 hours or less in oxygen following Paraquat injection.

Oxygen poisoning has many features which resemble paraquat intoxication. These similarities seem to be no mere coincidence. Early effects of oxygen poisoning at one atmosphere include injury to the lining of small blood vessels in the lungs (52) and formation of lung edema, probably by leakage from the injured blood vessels. Blood from

TABLE 1 -- 24 HOUR SURVIVAL OF RATS

	IN AIR	IN OXYGEN
Control	111/111	12/12
Paraquat (iv 27 mg/kg)	186/186	0/19
P	NS	0.001

Mortality of rats after intravenous injection of Paraquat 27 mg/kg. Breathing oxygen at one atmosphere greatly decreases survival.

oxygen-poisoned dogs interferes with the physical properties of dog lung surfactant (53). If the same is true in paraquat-intoxicated rats, then the inability to measure surfactant in lung tissue in these rats may be explained.

Paraquat is a non-inhaled environmental hazard which offers an unusual opportunity to piece together sufficient facts and reasonable hypotheses to account for much of the early clinical, physiological, and biochemical effects. In addition to its own toxic actions, it may interact with another environmental hazard and enhance susceptibility to oxygen poisoning of the lungs.

REFERENCES

1. Konwaler, B. F. (1950). Pulmonary emboli of cotton fibers. Amer. J. Clin. Path. 20: 385.

2. Wendt, V. F. et al. (1964). Angiothrombotic pulmonary hypertension in addicts. J. Amer. Med. Assoc. 188: 755.

3. Heinemann, H. O. and Fishman, A. P. (1969). Nonrespiratory functions of mammalian lung. Physiol. Rev. 49: 1.

4. Chinard, F. P. (1966). The permeability characteristics of the pulmonary blood-gas barrier. In: C. G. Caro, (ed.). Advances in Respiratory Physiology. Williams and Wilkens, Baltimore, 106.

5. Staub, N. C., Nagano, H. and Pearce, M. L. (1967). Pulmonary edema in dogs, especially the sequence of fluid accumulation in the lungs. J. Appl. Physiol. 22: 227.

6. Comroe, J. H. et al. (1953). Reflex and direct cardiopulmonary effects of 5-OH-tryptamine (serotonin). Amer. J. Physiol. 173: 379.

7. Louria, D. B., Hensle, T. and Rose, J. (1967). The major medical complications of heroin addiction. Ann. Int. Med. 67: 1.

8. Fisher, A. B. et al. (1870). Restoration of systemic blood flow to the lung after division of bronchial arteries. J. Appl. Physiol. 29: 839.

9. Gold, W. M. et al. (1965). Pulmonary-function abnormalities after lymphangiography. New Eng. J. Med. 273: 519.

10. Matthew, H. et al. (1958). Paraquat poisoning - lung transplantation. Brit. Med. J. 3: 759.

11. Graham, J. R. et al. (1966). Fibrotic disorders associated with methysergide therapy for headache. New Eng. J. Med. 274: 359.

12. Murray, M. J. and Kronenberg, R. (1965). Pulmonary reactions simulating cardiac pulmonary edema caused by nitrofurantoin. New Eng. J. Med. 273: 1185.

13. Granville-Grossmann, K. L. and Sergeant, H. G. S. (1960). Pulmonary edema due to salicylate intoxication. Lancet I: 575.

14. Heard, B. E. and Cooke, R. A. (1962). Busulphan lung. Thorax 23: 187.

15. Heard, B. E. (1962). Fibrous healing of old iatrogenic pulmonary edema ("hexamethonium lung"). J. Path. Bact. 83: 159.

16. Reichlin, S., Loveless, M. H. and Kane, E. G. (1953). Löeffler's syndrome following penicillin therapy. Ann. Int. Med. 38: 113.

17. Tuchman, H. (1954). Löeffler's reaction to para-amino salicylic acid. Amer. Rev. Tuberc. 70: 171.

18. Rodman, T., Fraimow, W. and Myerson, R. M. (1958). Löeffler's syndrome: report of case associated with administration of mephenesin carbamate (Tolseram). Ann. Int. Med. 48: 668.

19. Clarysse, A. M. et al. (1969). Pulmonary disease complicating intermittent therapy with methotrexate. J. Amer. Med. Assoc. 209: 1861.

20. Siegel, M., Lee, S. L. and Peress, N. S. (1967). The epidemiology of drug-induced systemic lupus erythematosus. Arth. Rheum. 10: 407.

21. Green, G. M. and Kass, E. H. (1964). Factors influencing the clearance of bacteria by the lung. J. Clin. Invest. 43: 769.

22. Germuth, F. G. (1956). The role of adrenocortical steroids in infection, immunity, and hypersensitivity. Pharmacol. Rev. 8: 1.

23. Reed, D. L. and Coon, W. W. (1963). Thromboembolism in patients receiving progestational drugs. New Eng. J. Med. 269: 622.

24. Weill, H. et al. (1964). Early lipoid pneumonia. Amer. J. Med. 36: 370.

25. Waring, J. I. (1933). Pneumonia in kerosene poisoning. Amer. J. Med. Sci. 185: 325.

26. Brünner, S., Rovsing, H. and Wulf, H. (1964). Roentgenographic changes in the lungs of children with kerosene poisoning. Amer. Rev. Resp. Dis. 89: 250.

27. Spencer, H. (1968). Pathology of the Lung. (2nd ed.) Pergamon Press, London, 479.

28. Warren, S. (1942). Effects of radiation on normal tissues. Arch. Path. 34: 917.

29. Whitfield, A. G. W., Bond, W. H. and Arnott, W. M. (1956). Radiation reactions in the lung. Quart. J. Med. 25: 67.

30. Adamson, I. Y. R., Bowden, D. H. and Wyatt, J. P. (1970). A pathway to pulmonary fibrosis: An ultrastructural study of mouse and rat following radiation to the whole body and hemithorax. Amer. J. Path. 58: 481.

31. Kohen, H. and Biskind, G. R. (1946). Pathologic aspects of atmospheric blast injuries in man. Arch. Path. 42: 12.

32. Zuckerman, S. (1940). Experimental study of blast injuries to the lungs. Lancet II: 219.

33. Andersen, P. and Loken, S. (1968). Lung damage and lethality by underwater detonations. Acta Physiol. Scand. 72: 6.

34. Visscher, M. B., Haddy, F. J. and Stephens, G. (1956). The physiology and pharmacology of lung edema. Pharmacol. Rev. 8: 389.

35. Valdivia, E. and Sonnad, J. (1966). Fatty change of the granular pneumocyte in CCl_4 intoxication. Arch. Path. 81: 514.

36. Gould, V. E. and Smuckler, E. A. (1970). Acute alveolar injury in experimental carbon tetrachloride intoxication. Presented in part at the Hanford Biology Symposium on Air Pollution and Lung Biochemistry, Richland, Washington.

37. Spiezer, F. E. and Frank, N. R. (1966). The uptake and release of SO_2 by the human nose. Arch. Env. Health 12: 725.

38. Frank, N. R. et al. (1967). The diffusion of 35-SO_2 from tissue fluids into the lungs following exposure of dogs to 35-SO_2. Health Physics 13: 31.

39. Kotin, P. and Falk, H. L. (1967). Atmospheric factors in pathogenesis of lung cancer. Adv. Cancer Res. 7: 500.

40. Bullivant, C. M. (1966). Accidental poisoning by Paraquat: Report of two cases in man. Brit. Med. J. 1: 1273.

41. Almog, C. H. and Tal, E. (1967). Death from Paraquat after subcutaneous injection. Brit. Med. J. 3: 721.

42. Oreopoulos, D. G. et al. (1968). Acute renal failure in case of Paraquat poisoning. Brit. Med. J. 1: 749.

43. Duffy, B. S. and O'Sullivan, D. J. (1968). Paraquat poisoning. J. Irish Med. Assoc. 61: 97.

44. Fennelly, J. J., Gallagher, J. T. and Carroll, R. J. (1968). Paraquat poisoning in a pregnant woman. Brit. Med. J. 3: 722.

45. Campbell, S. (1968). Death from Paraquat in a child. Lancet I: 133.

46. Hargreave, T. B., Gresham, G. A. and Karayannopoulos, S. (1969). Paraquat poisoning. Postgrad. Med. J. 45: 633.

47. Mantkelow, B. W. (1967). The loss of pulmonary surfactant in Paraquat poisoning. Brit. J. Exp. Path. 48: 366.

48. Fisher, H. K. and Clements, J. A. (1969). Effects of the herbicide paraquat on mechanical properties or rat lungs. Fed. Proc. 28: 526.

49. Clements, J. A., Mellenbogen, J. and Trahan, H. J. (1970). Pulmonary surfactant and evolution of the lungs. Science 169: 603.

50. Davidson, C. L. and Papirmeister, B. (1971). Bacteriostasis of E. coli by the herbicide Paraquat. Proc. Soc. Exp. Biol. Med. 136: 359.

51. Fisher, H. K., Clements, J. A. and Tierney, D. F. (1970). Early pulmonary effects of Paraquat in rats. Clin. Res. 18: 190.

52. Kistler, G. S., Caldwell, P. R. B. and Weibel, E. R. (1967). Development of fine structural damage to alveolar and capillary lining cells in oxygen-poisoned rat lungs. J. Cell Biol. 33: 605.

53. Caldwell, P. R. B. et al. (1965). Effects of oxygen breathing at one atmosphere on the surface activity of lung extracts in dogs. Ann. N. Y. Acad. Sci. 121: 823.

CHAPTER 8. SOME SPECIFIC EFFECTS OF ENVIRONMENTAL AGENTS

BENJAMIN G. FERRIS, Department of Physiology,
Harvard School of Public Health,
Boston, Massachusetts

There are rather few specific reactions of the lung that can be attributed to a particular material or agent. This results, in part, from the limited responses that the lung can make to a variety of different stimuli. Examples of specific responses are the granulomata produced by exposure to beryllium, the nodules of silicosis, and the primary lesion of tuberculosis. More commonly, an observed effect is the result of a number of different stimuli. For example, asthmatic attacks or bronchial constriction can be triggered by inhaling a wide variety of irritant gases such as SO_2, Los Angeles "Smog" (1), cigarette smoke (2) or inert dust (3), although the mechanism of action is not clear. In the case of SO_2, it is thought to be a reflext response secondary to the absorption in the upper air ways. The action of cigarette smoke may either be due to direct action of certain components or due to the release of histamine in the lung.

An increased frequency of lower respiratory tract disease has been reported to be associated with increasing levels of SO_2 in the ambient air (4), but apparently not with other pollutants. A similar study failed to show any effect on upper respiratory tract disease (5). The greater sensitivity of the lower respiratory tract has also been demonstrated by studies in children (6-8). In some studies, decreased pulmonary function has been identified in children (6,8,9), and adults (10,11). The medical significance of these statistically significant changes has not been clearly demonstrated; intuitively, one would think that they are medically significant. Decreases in pulmonary function occur with age in the adult. Whether these changes represent "aging" per se, or the end result of the insults of airborne materials, is not apparent. The observation that cigarette smoking seems to enhance this effect (12) could be interpreted to demonstrate that aging really is the result of repeated exposure of the lung to air pollutants. Whether the pollutants are directly responsible for tissue destruction, whether they alter structures which are then more susceptible to damage by the normal stresses and strains, or whether the pollutants in some way interfere with the normal cleansing activity within the lung such as ciliary action, formation of mucus, or macrophage activity, is not known.

Studies on animals have shown that a number of irritants can interfere with ciliary action (13) and mucus flow (14). These observations have been extended to man (15), in whom cigarette smoke has been shown to have variable effects depending upon the characteristics of the individual. In general, however, cigarette smokers or miners had poorer lung clearance of radioactive particles than non-exposed individuals. Poor clearance may lead to an accumulation of mucus, bacteria, or particulates in the lung. This, in turn, could result in the symptoms of chronic bronchitis and eventual lung damage. In coal workers' pneumoconiosis, for example, considerable scarring, tissue destruction, and chronic bronchitis develop despite the older belief that coal is non-fibrogenic. It may well be that the heavy load of coal dust to be cleared interferes with the removal of other more noxious compounds such as silica dust, or bacteria. The centrilobular emphysema that develops in this condition could reflect such action although the exact mechanism is not known.

In some exposures, as to cotton dust, histamine is released in the lung. This in turn, produces constriction of the alveolar ducts and respiratory bronchioles. Here, too, chronic disabling lung disease can develop, apparently as a result of these repeated episodes of constriction that somehow result in permanent lung damage and destruction of parenchyma. Similar responses occur in exposures to toluene diisocyanate (TDI), bagasse (fibre residue of crushed sugar cane) and mouldy hay. These latter conditions are the result of immunologic responses to the inhaled material. Interference with lung clearance may also result if the alveolar macrophages are not able to function properly. In vitro studies (16) have indicated that the alveolar macrophages are sensitive to SO_2, oxides of nitrogen, and ozone. The levels of SO_2 and oxides of nitrogen that produced an effect were relatively high. The ozone was particularly toxic at 0.06-0.09 ppm. These are levels of ozone that are common in Los Angeles. Whether similar results occur in vivo is not known. Some components of air pollution from coal burning sources can enhance the growth of H. influenzae in cultures (17) although it has not been demonstrated in vivo. On the other hand, the most common organism cultured from sputum in Great Britain is H. influenzae. Thus, the possibility exists that air pollutants may interfere with activity of macrophages and, either by this means or by direct action, enhance the growth of pathogenic organisms. Studies in animals have shown that pre-exposures to various irritant gases can enhance susceptibility to infection (18).

To assess the effects on intact man, various tests of pulmonary function have been done on population groups for chronic exposures, or on volunteers for short-term exposures, at relatively higher levels. Working populations have also been studied, but the extrapolation of findings from such selected groups must be done with caution. The working population is always a selected population, sometimes highly so, consisting of those who can tolerate the prevailing conditions. The general population, on the other hand, contains many elderly, infirm, or particularly sensitive individuals who can be expected to react more readily or severely. A failure to detect changes as a result of exposure at work does not mean that no effect is to be expected in general populations. Conversely, if an effect is seen in the working population, one can expect a greater effect for comparable exposures in the general population. For example, in a study of men in a pulp and paper mill, little effect from SO_2 was seen in the workers exposed, whereas exposures to Cl_2 did produce changes in the flow-volume curves of the men exposed to it (19). From such observations, one could say that Cl_2 was more damaging than SO_2, but one should not predict that SO_2 would be without harm to general populations.

Studies over different levels of air pollution (SO_2 and particulates) that are reasonably comparable, and that have been controlled for cigarette smoking, have shown an increased prevalence of respiratory symptoms such as phlegm production, wheezing in the chest, or exacerbations of chest illness in adult populations (10,20). Some of these symptoms, such as phlegm production, may be more related to the particulate levels than to the SO_2 (21,22); that is, when levels of SO_2 and suspended particulates have been high, there are more symptoms, and conversely, when the levels of pollution are low, symptoms are less. In these studies, carried out in London, phlegm production or exacerbations of symptoms decreased despite continued high levels of SO_2 as the particulates were markedly reduced. It is not known whether there is interaction between the particulates and SO_2 to produce a more toxic chemical, whether the SO_2 is simply adsorbed onto the particle, whether the adsorbed SO_2 in the presence of moisture in the respiratory tract is converted to H_2SO_3 or H_2SO_4, or whether some other unrecognized chemical reactions take place.

The effect of cigarette smoking on respiratory symptoms has been well documented in the Surgeon General's Report on Smoking (23). Since that report, other studies have shown that cigarette smoking causes a more rapid decline in forced vital capacity ($FEV_{1.0}$) and peak expiratory flow rate (PEFR) with age (12), that college students who smoke have altered flow-volume curves (24), and that the transfer factor for CO is decreased (25). Ex-smokers tend to be more like non-smokers, so that there appears to be a reversible element in the change. Whether it is in the membrane component or in the volume of blood in the capillaries is still not clear. One study demonstrated more change in the membrane component (26) and another in the volume of blood in the capillaries (27) in association with cigarette smoking. Some of this discrepancy may reflect the different populations studied. In the first study, subjects were selected who had no respiratory symptoms or history of chronic respiratory disease. In the second study, a random sample of adults from a community was studied. Thus, cigarette smoking is a major factor in chronic bronchitis and probably chronic obstructive lung disease. Its role as a major factor in causing cancer of the lung is also well documented (23). (See also Chapters 3 and 12).

Studies on the oxidant type of pollution have shown some interesting results. There is little evidence that oxidant pollution is a factor in the causation of chronic bronchitis, despite the obvious irritating quality of the pollution to the mucous membranes of the eye, nose, and throat. Persons with pre-existing chronic lung disease, however, do show changes in their pulmonary resistance when they are exposed to air polluted with oxidants (28). Acute exposures of volunteers to 1 ppm ozone have demonstrated a decreased transfer factor for CO (29), and athletes doing long distance running have shown poorer times (i.e. poorer performance) during periods of increased oxidant (30). Whether this is a reflection of the decreased transfer factor or some other parameter, such as irritation of the mucous membranes, is not known. Studies in the laboratory on animals have shown that exercise exacerbates the effect of ozone, and that a tolerance to the material can be developed (31). Again, we do not know the relative effect or importance of these responses in man, or whether the development of tolerance can explain some of the lack of response of the Los Angeles population to oxidant pollution.

Studies in bronchitic patients have shown alterations 'in dynamic compliance of the lung, increased alveolar-arterial oxygen differences, increased dead space-tidal volume ratios, and decreased steady-state diffusing capacity (32). These reflect ventilation-perfusion abnormalities. It is not possible to separate out whether the abnormality lies more with ventilation or with perfusion, or which change occurred first. In any case, there is destruction of lung parenchyma, either by direct action or by interference with blood supply. Pathologic studies on human population in two areas with different levels of air pollution have shown increased amounts of parenchymal tissue destruction in the area with higher air pollution (33). Cigarette smoking habits were taken into account. Thus, a variety of inhaled materials can produce lung tissue damage and altered pulmonary function.

Irritant gases such as SO_2, NO_2, and Cl_2 can cause direct damage to pulmonary tissue. This may produce pulmonary edema, which in turn may be absorbed without residual effect or, if more severe, may lead to scarring and fibrosis within the lung. In the case of SO_2, relatively high concentrations must be inhaled, sufficient to overwhelm the absorptive capacity of the upper airways. With NO_2 and Cl_2 much lower concentrations are required to produce the damage. These differences in concentration needed to cause damage reflect the different water solubilities of the gases. Bronchiolitis obliterans has occurred as a result of exposures to NO_2.

We should not overlook the potential for carcinogenesis. Cigarette smoking has been indicted (23); chromates, arsenic, asbestos, radio-active particles, and certain organic compounds can cause pulmonary cancers. Nickel carbonyl $Ni(CO)_4$ has also been shown to be a nasal and pulmonary carcinogen as have certain woods (34). There may also be interaction between materials such as cigarette smoke and asbestos or uranium in the case of cancer, or coal dust and cigarette smoke in coal miners' pneumoconiosis.

REFERENCES

1. Schoettlin, C. E. and Landau, E. (1961). Air pollution and asthmatic attacks in the Los Angeles area. Public Health Reports, 76: 545.

2. Nadel, J. A. and Comroe, J. H. (1961). Acute effects of inhalation of cigarette smoke on airway conductance. J. Appl. Physiol. 16: 713.

3. DuBois, A. and Dautrebande, L. (1958). Acute effects of breathing inert dust particles and of carbachol aerosol on the mechanical characteristics of the lungs in man. Changes in response after inhaling sympathomimetic aerosols. J. Clin. Invest. 38: 1746.

4. Dohan, F. C. and Taylor, E. W. (1960). Air pollutants and incidence of respiratory disease. A preliminary report. Amer. J. Med. Sci. 240: 337.

5. Ipsen, J. (1965). Relationship of acute respiratory disease to measurements of atmospheric pollution and local meterological conditions. Final report. Henry Phipps Inst. Contract No. PH 86-63-25.

6. Lunn, J. F., Knowelden, J. and Handyside, A. J. (1967). Patterns of respiratory illness in Sheffield infant school children. Brit. J. Prev. and Soc. Med. 21: 7.

7. Douglas, J. W. B. and Waller, R. E. (1966). Air pollution and respiratory infection in children. Brit. J. Prev. and Soc. Med. 20: 1.

8. Biersteker, K. and Van Leeuween, P. (1970). Air pollution and peak flow rates of school children in two districts in Rotterdam. Arch. Env. Health 20: 382.

9. Ferris, B. G. (1970). Effect of air pollution on school absences and differences in lung function in first and second graders in Berlin, New Hampshire. Jan. 1966 to June 1967. Amer. Rev. Resp. Dis. 102: 591.

10. Ferris, B. G. and Anderson, D. O. (1964). Epidemiological studies related to air pollution. Comparison of Berlin, New Hampshire and Chilliwack, British Columbia. Proc. Roy. Soc. Med. (Suppl.) 57: 979.

11. Holland, W. W. et al. (1965). Respiratory disease in England and the United States. Studies of comparative prevalence. Arch. Env. Health 10: 338.

12. Ferris, B. G., Anderson, D. O. and Zikmantel, R. (1965). Prediction values for screening tests of pulmonary function. Amer. Rev. Resp. Dis. 91: 252.

13. Dahlman, T. (1956). Mucous flow and ciliary activity in the trachea of healthy rats and rats exposed to respiratory irritant gases (SO_2, NH_3 HCHO). Acta Physiol. Scand. 36: 1. (Suppl. 123).

14. Spiegelman, J. R. et al. (1968). The effect of acute SO_2 exposure on bronchial clearance in the donkey. Arch. Env. Health 17: 321.

15. Albert, R. E., Lippman, M. and Briscoe, W. (1969). The characteristics of bronchial clearance in humans and the effects of cigarette smoking. Arch. Env. Health 18: 738.

16. Weissbecker, L. et al. (1969). In vitro alveolar macrophage viability. Arch. Env. Health 18: 756.

17. Lawther, P. J., Emerson, T. R. and O'Grady, F. W. (1969). H. influenzae growth stimulation by atmospheric pollutants. Brit. J. Dis. Chest 63: 45.

18. Coffin, D. L. (1970). The relationship of infectious agents and air pollutants. Interregional symposium on air quality criteria and guides. WHO, Geneva, Working paper No. 16.

19. Ferris, B. G., Burgess, W. A. and Worcester, J. (1967). Prevalence of chronic respiratory disease in a pulp mill and a paper mill in the United States. Brit. J. Ind. Med. 24: 26.

20. Reid, D. D. et al. (1964). An Anglo-American comparison of the prevalence of bronchitis. Brit. Med. J. 2: 1487.

21. Speizer, F. E. (1969). An epidemiological appraisal of the effect of ambient air on health. Particulates and oxides of sulfur. J. Air Pollut. Control Assoc. 19: 647.

22. Lawther, P. J., Waller, R. E. and Henderson, M. (1970). Air pollution and exacerbations of bronchitis. Thorax 25: 525.

23. U. S. Department of Health, Education, and Welfare. (1964). Smoking and Health. Report of the advisory committe to the surgeon general of the Public Health Service. PHS Pub. 1103.

24. Peters, J. M. and Ferris, B. G. (1967). Smoking, pulmonary function, and respiratory symptoms in a college age group. Amer. Rev. Resp. Dis. 95: 775.

25. Cotes, J. E. and Hall, A. M. (1970). The transfer factor for the lung; normal values in adults. In: P. Arcangeli et al., (eds.). Introduction to the Definition on Normal Values for Respiratory Function in Man. Panminerva Medica, 338.

26. Wilson, R. M. et al. (1960). The pulmonary pathologic physiology of persons who smoke cigarettes. New Eng. J. Med. 262: 956.

27. Van Ganse, W. F., Ferris, B. G. and Cotes, J. E. (1971). Cigarette smoking and pulmonary diffusing capacity (transfer factor). Amer. Rev. Resp. Dis. (In press).

28. Motley, J. L., Smart, R. H. and Leftwich, C. I. (1959). Effects of polluted Los Angeles air (smog) on lung volume measurements. J. Amer. Med. Assoc. 171: 1469.

29. Young, W. A., Shaw, D. B. and Bates, D. V. (1964). Effects of low concentrations of ozone on pulmonary function. J. Appl. Physiol. 19: 765.

30. Wayne, W. S., Wehrle, P. F. and Carroll, R. E. (1967). Oxidant air pollution and athletic performance. J. Amer. Med. Assoc. 199: 901.

31. Stokinger, H. E., Wagner, W. D. and Wright, P. G. (1956). Studies in ozone toxicity potentiating effects of exercise and tolerance development. Arch. Indust. Health 14: 158.

32. Levine, G. et al. (1970). Gas exchange
     abnormalities in mild bronchitis and
     asymptomatic asthma. New Eng. J. Med. 282:
     1277.

33. Ishikawa, S. et al. (1969). The "Emphysema
     Profile" in two midwestern cities in North
     America. Arch. Env. Health 18: 660.

34. Acheson, E. D. et al. (1968). Nasal cancer
     in wood workers in the furniture industry.
     Brit. Med. J. 1: 587.

CHAPTER 9. ENVIRONMENTAL FACTORS IN CHRONIC LUNG DISEASE

JOHN P. WYATT, University of Manitoba,
School of Medicine, Winnipeg, Canada

> "This most excellent canopy, the air ....
> why it appears no other thing to me than a
> foul and pestilent congregation of
> vapours."
>
> Hamlet, Act II, Scene II.

Twelve years after the first production of Hamlet, John Evelyn in 1616 attempted to relate chronic respiratory disease to air pollution (1). In his essay "Smoake of London", he writes about "the hellish and dismal cloud of 'sea coale' -- impure and thick mist accompanied with a fuliginous and filthy vapor .... corrupting the lungs .... so that Catharrs, Phthisicks, Coughs, and Consumptions range more in this one city than in the whole earth besides". This is the ultimate in simplistic views: Bad air -- diseased lungs! But Evelyn's essay was the herald for our contemporary concern on pollution induced pulmonary disease.

Owing to the complex nature of air contaminants which make up Shakespeare's "congregation of vapors", it is difficult to indict any specific agent as being solely responsible for this loosely defined group of non-specific chronic lung diseases in man. A review of the evidence indicates that, even in the dramatic air pollution catastrophes of the past, which developed in the heavily industrialized valleys of the Meuse in Belgium (2) and the Monongahela in Pennsylvania (3), or the four days of black fog in London, 1952 (4), no single agent was incriminated as being directly responsible for the attendant high incidence of pulmonary disease. One conclusion was inescapable: the excessive mortality was the result of irritation of the respiratory tract by air contaminants.

The second cardinal conclusion that emerges from these air pollution respiratory calamities is the knowledge that chronic cardiopulmonary disease was present in many of the victims. These air pollution episodes led to an immediate increase in respiratory disease mortality and possibly one more event in fostering an endemic chronic disease. Although there is little a priori knowledge or a posteriori evidence to guide us, it is tempting to propose that these victims had developed their chronic pulmonary disease earlier owing to their exposure to the ambient contaminants during years of residence in these air-insulted regions. In most of these acute pulmonary disease disasters of the past, the victims were in the older age groups. In those of younger years, the defense mechanisms of the lung, relatively uncompromised, may have been able to neutralize the injurious actions of the acute episode, but they are ineffective against prolonged and insidious exposures and actions of air contaminants. Although this premise appears, in part, to be derived by sophistry, it offers an opportunity to survey the extent to which past and current data support this extrapolation on environmental lung pathology.

CHRONIC LUNG DISEASE IN ADULTS

The distinctive mechanisms and range of causal agents capable of producing 'specific' lung injury are discussed in greater detail in other chapters in this monograph. Based on direct and collateral evidence, the two commonest long standing lung alterations that have a relationship to community pollution, and of themselves may be interrelated, are chronic 'interstitial' pneumonitis and the respiratory disease complex, bronchitis-emphysema. The chronic pneumonitis group with reparative fibrosis is, in general, a proliferative cellular response; the bronchitis-emphysema complex is one initially of irritation but eventually irreparable loss of sustentacular tissue components.

From the standpoint of etiology, the best evidence linking the proliferative cellular response in man to air contaminants rests on experimental investigation -- known pollutants in measured doses producing a recognizable histologic picture. The erosive-dissolutive portrait of bronchitis-emphysema, on the other hand, has limited evidential support from the laboratory animal studies. The influence of air pollutants in the development of bronchitis-emphysema rests on the evaluation of the human lung and its behavior in an altered external environment.

Experimental Evidence for Chronic Proliferative Lung Disease

Various laboratory observations support the idea that urban air contaminants, such as sulfur dioxide, photochemical oxidants, oxides of nitrogen, and various hydrocarbons can be shown to affect adversely the various protective mechanisms of the respiratory tract, such as ciliary movement, mucous production, and pulmonary clearance of infectious disease agents.

A popular chemical agent in the hands of the experimentalist has been sulphur dioxide (5,6,7). It has been repeatedly demonstrated that this agent has a uniform and titratable effect on the airway over long periods of exposure. Inhalation of sulphur dioxide in experimental animals results in temporary spasm of the smooth muscle of the bronchioles; somewhat higher concentrations cause increased mucus production on the walls of the upper airways; still higher concentrations result in severe inflammatory responses in the mucosa, associated with desquamation of the bronchiolar surface epithelium and intra-luminal blockage. For the production of lesions of a chronic nature, Lamb and Reid (8) exposed 48 rats (bred from pathogen-free animals) to SO_2 at 400 ppm for a period of three hours a day, five days a week up to 35 days, and obtained hypertrophy of the bronchial mucous glands, and a proliferation of goblet cells which extended to the most peripheral points of the respiratory bronchioles. These two pathologic alterations, mucous gland hypertrophy and goblet cell proliferation, are the two key lesions in the chronic bronchitis of man (9).

In the early evolution of nitrogen dioxide injury, Wagner and co-workers (10) found a transient increase in oxygen consumption in rabbits exposed intermittently to 25 ppm of nitrogen dioxide. In general, a persistent increase in pulmonary resistance with bronchiolar lesions (11) can be produced in the particularly susceptible guinea pig (12). The data garnered by Davidson et al (13) reinforces the above data. Adult rabbits were continuously exposed to 8-12 ppm of NO_2, for approximately three months; the residual volume increased as well as the non-elastic resistance, and arterial hypoxia was recorded. On removal of rabbits from the exposure chambers and return to room air, the physiologic changes were completely reversed, but pathologic alterations were reported on the scarified animals "as an abnormal enlargement of the distal air spaces."

Animals subjected for a period of months to daily inhalations of sublethal doses of ozone develop fibrotic thickening of the walls of their bronchioles (14). It is also known that ozone inhalations, at levels of concentrations significantly below those inducing fatal pulmonary edema, can produce the late manifestations of pulmonary resistance (15).

Studies of the ambient atmosphere of some industrialized cities on biologic tissues have been inconclusive; with the record of only minimal septal wall inflammatory changes in rats directly exposed to the complexes of the polluted air. Garner et al (16) studied pulmonary changes in 7,000 mice after prolonged exposure to ambient polluted Los Angeles air. In several strains of mice, it was demonstrated that such an exposure lead to an increased susceptibility for pulmonary infections.

Although an exact replication of chronic pulmonary disease is difficult to produce consistently in the animal laboratory, certain common pathological patterns emerge. One irritant may have a greater effect upon one pulmonary defence mechanism, a second agent injures another pathway of defence; both produce a chronic non-specific lung response (17).

General Commentary on the Proliferative Response

The progressive proliferative and fibroblastic responses seen in the experimental animal, due to a variety of agents from the oxides of nitrogen though ozonides to hydrocarbons, may be a sequel of direct cell damage or it may be related to enhancement of bacterial activity in the lung. The injurious agents may be liberated or formed in the ambient atmosphere and, with their repeated lodgement in terminal bronchioles, may produce progressive harmful effects on previously impaired bronchial mucosa. Fine particulate agents in the neighborhood of 2μ impinge upon the respiratory membrane, and the development of a chronic pneumonitis becomes an anticipated event, with the dominant degree of cellularity being within the septa (interstitial), or associated with an intermittent or a continuous desquamation of alveolar cells into the distal respiratory unit.

Of the many environmental agents (18) which have been labelled fibrogenic, a few, such as beryllium, manganese and radioactive compounds produce tissue responses of a granulomatous or necrotizing nature. But, in general, the fluid and cell responses to pulmonary irritants follow a common pathway. Apart from acute edemagenic responses (19) in protracted low dose exposures, the tissue response is often of an undistinguished bronchiolar and/or septal 'pneumonitis' pattern. There appears to be a relative uniformity in the pathologic response attributable to different air pollutants. Any interference with pulmonary defence mechanisms leads to injury, and the reparative processes that follow may leave the lung vulnerable to the same or other noxious agents. Eventually the critical reserve of the respiratory system may be consumed.

In addition to the direct action of agents damaging the respiratory system or encouraging of microbiologic infections, immunologic disorders may follow the inhalation of environmental antigenic challenges and lead to chronic lung disease. In the large group of type III lung hypersensitivity reactions, morphologically labelled extrinsic allergic alveolitis (20,21), byssinosis probably offers a clinical and pathologic model of a progressive lung disorder attributable to a known pollutant (22). The weekly repetition of Monday morning fever and tightness in the chest has been known for years in the cotton industry in the Midlands of England. This progressive pulmonary disability has been assigned to a specific industrial pollutant, but it may be worsened by non-specific air pollutants (23). The findings in long standing cases revealed a non-specific septal wall thickening of a mixed cellular and fibroblastic nature, and pulmonary heart disease was a frequent companion.

Chronic Bronchitis--Emphysema Complex

These two pathologic responses are considered together, as in most cases they are coexistent, although from an etiologic viewpoint they may have distinctive differences in their origins. The chronic bronchitis component, unlike the necrotizing bronchiolitides of adenovirus infections or acute experimental injuries, is usually associated with some degree of lung dissolution. This latter morphological change of lung erosion and destruction represents the emphysematous portion of the complex. Although a number of pathogenic explanations have been proposed for the progressive dissolution of the distal respiratory units (emphysema), the mechanism of this lung destruction is, as yet, unclear (9).

Smoking has been blamed by many scientists as the sole etiologic factor in chronic bronchitis and emphysema. Although the incrimination of this factor, owing to its strong associative role from clinical, physiologic, and pathologic studies, may dominate scientific investigations, other factors are by no means eliminated from having a primary as well as a derivative role. Evidence has steadily accumulated emphasizing the multi-factorial background in chronic bronchitis and emphysema (24,25).

There is ample support to buttress a relationship between the inhalation of various dusts and the incidence of respiratory infections (17). The bactericidal activity of the lung is decreased by exposure to coal dust, and the efficiency of the muco-ciliary escalator is seriously impaired by ciliotoxic agents such as the oxides of sulphur or nitrogen. On the basis of the numerous factors that may induce and perpetuate a disturbance of the mucus-ciliary clearance mechanism, chronic bronchitis is inevitably a combined clinical and pathologic entity; implicit in this diagnosis is the role of industrial and urban pollutants.

The opportunity to follow the entire gamut of pathologic alterations that one sees, for instance, in 'respirator lungs' is rarely presented in studies of community pollution (26). In the last ten years there has been a plethora of publications dealing with chronic non-specific lung disease, in particular bronchitis-emphysema, but the majority of these writings have dealt ad nauseam with structure-functional relationships, the linkage with cigarette smoking and, the morphologic characterization of the lung damage.

The bulk of evidence linking the etiology of the complex with air pollutants has rested largely on clinical epidemiologic studies. Proof of chronic disease causation is frequently impossible. To establish the significance of a pollution-effect relationship, it is necessary to demonstrate an excess risk, a dose-response relationship, constancy and consistency of the relationship and, if possible, the production of similar effects in susceptible experimental animals.

Epidemiological Evidence on the Role of Air Pollution in Chronic Lung Disease

How much of the increase in chronic pulmonary disease may be related to environmental pollution is difficult to determine, but it has certainly occurred primarily in major urban areas, paralleling increasing industrialization, automobile usage, and other "community" factors. Many investigators use air pollution and the chemical complexes which contribute to contaminated air as the major surrogate in any "community factor" study. It is not surprising, therefore, to see a number of epidemiological reports which repeatedly confirm that the prevalence of bronchitis and chronic respiratory disease mortality is associated with: (1) the amount of fuel consumed, (2) levels of annual sulphur dioxide measurement in the ambient air, (3) levels of both settled and air-borne dust.

Chronic bronchitis-emphysema is our most rapidly rising cause of death, the rate having increased more than tenfold during the past 15 years. The popular explanation for this alarming increase in the prevalence of complex bronchitis and emphysema has been that personal pollution (cigarette smoking) is solely responsible. This explanation fails to take into consideration the role of air contaminants known to be irritative to respiratory membranes. An arrest of incidence, or even a decline in the morbidity and mortality curves, of bronchitis-emphysema in the United Kingdom has been noted since the introduction of the Clean Air Act in Great Britain in 1965. This recent finding re-emphasizes the view that a continuing analysis of the role of air pollutants in chronic non-specific lung disease is essential, particularly as the background of environmental contamination is under constant change due to air pollution control measures.

Epidemiological data are the kind of health data best adapted to the estimation of air pollution effects. Although there are many caveats that enter into any prospective, and even more so into any retrospective analysis, of air pollution and its pulmonary effects, the general magnitude of the association cannot be doubted. Although the investigation of the morbidity ratio is usually more fruitful to the epidemiologist seeking associations, in this essay we have concentrated on mortality rates, particularly if they are attributable to established pathological reactions such as asthma, chronic bronchitis, and emphysema.

Established cases of bronchial asthma appear to behave as the litmus of lung reactivity in response to air pollution. Numerous studies in large industrialized cities have revealed, on the average, a ten-fold greater incidence of bronchial asthma during days of stagnant and high air pollution than on days with cleaner air. Other studies have correlated the incidence of asthmatic attacks among adults with intensities of pollution. Although there may be real differences in the prevalence of bronchitis-emphysema (recurrent disabling chest infections) between England and United States, transatlantic cooperative studies indicate many agreements, and certainly communication for comparison purposes has improved. With a similar definition for chronic bronchitis in Denmark as well as the United Kingdom, the rate of bronchitis in Denmark, an agrarian country, was one-sixth that of Great Britain (27). In Denmark there was no difference in the bronchitis rate between rural and urban areas. In Britain, men between the ages of 45 and 54 have five times the death rate for certain chronic respiratory diseases, as compared with similar age groups in North America. Similarly, a higher death rate due to chronic respiratory diseases has been demonstrated in Canadian veterans who lived five or more years in-a city. This increased incidence of chronic respiratory disease is obviously predominant in the urban areas and finds a similar reflection on a broad geographic canvas made up of many lands such as Norway (28), England (29), U.S.A., (30) and Japan (31). Epidemiologic studies, in addition to demonstrating the predominance of chronic respiratory disease, also indicate a relationship with oxidizing forms of pollution such as that in Los Angeles, or a high correlation with the annual average concentration of sulphur dioxide as, for example, in Genoa, Italy.

In a recent extensive review of the literature dealing with air pollution and respiratory disease, Lave and Seskin (32), by multiple regression analyses of the large number of epidemiologic statistics published earlier, concluded that the significant explanatory variable for all fatal cases of bronchitis was air pollution. These Pittsburgh investigators calculated from the Buffalo, N. Y., statistics, for instance, that the mortality for asthma, bronchitis, and emphysema could be reduced 50% if the air in all of Buffalo were cleansed to equal the air in those areas that had the best air. The effect of chronic bronchitis and emphysema upon the right heart is well known; utilizing pulmonary heart disease as the parameter, Kasper has reported a two fold higher frequency of cor pulmonale in heavily polluted districts in Czechoslovakia (33).

The difficulties in assessing chronic respiratory disease in communities where the differences between intensity of pollution may be slight, was first reflected by the carefully controlled clinical studies in the Seward-Florence township study (32). This investigation, as well as those in the two other town models of Berlin, N.H. (34) and Chilliwack, B.C. (35) emphasized that air pollution studies, the effect of selective migration, and occupational and personal air pollution, as well as socioeconomic factors, may restrict the value of the scientific data extracted. For that reason, and taking into consideration as much as possible the statistical lessons learned from the classical epidemiologic investigations cited above, we inserted into our ecological study an additional strengthening factor -- the objective and quantitative evidence of anatomic lung destruction (36).

Three hundred lungs in a retrospective autopsy series involving two cities, St. Louis, Missouri and Winnipeg, Canada, were analyzed as to the extent and degree of emphysema by a standardized grid method measurement (9). These two cities have striking differences in their geographic, industrial, and meteorologic profiles. St. Louis, a heavily industrialized city, is in the Mississippi-Missouri Valley often covered by a pall of warm smog for most of the year. Winnipeg has limited or light industry, being primarily the centre of an agricultural area and a trans-shipping centre; the air is dry, the terrain is flat and there is a 12 mph prevailing wind from Hudson Bay southwards. The urban groups were matched for smoking habits, length of residence in the city, age at immigration, and employment history ("dusty" occupations were excluded).

The immigrants in both city populations were all from Europe, had come to North America before the age of 30 years, and had lived two-thirds of their lives in the same geographic area. Heavy smokers were

TABLE 1 -- STUDY POPULATION FOR THE 'EMPHYSEMA PROFILE'

	CASES	MEAN AGE	FEMALE	%	IMMIGRANT MEAN AGE MIGRATED	NON-SMOKERS %
ST. LOUIS	300	60.4	96	9	22	25
WINNIPEG	300	59.2	93	20	28	26

defined as smoking 30 cigarettes a day for at least 25 years.

Although a lifetime exposure to pollutants might bear little relationship to currently measured levels, St. Louis had, as far back as 1941, suppressed the burning of "dirty" soft coal and largely prevented this additive contaminant from entering into the metropolitan environs.

The anatomic evidence of emphysema was graded into mild (less than 20% of lung destroyed); and advanced and severe (over 70%); with moderate emphysema in between.

Analysis revealed a much greater prevalence of autopsy proven emphysema in the American city. Very few cases of emphysema were observed in Winnipeg under fifty years of age; only in the old age group did the curves of prevalence of emphysema approach one another. Among the non-smokers, there was more emphysema in the St. Louis group than in the Winnipeg group, and in this latter group the degree of severity of the emphysema was much less. The charts reveal that pulmonary emphysema had an earlier age distribution and was much more severe in St. Louis males than in males resident for a long time in Winnipeg. The incidence of severe emphysema in male smokers was four times as high in St. Louis than it was in Winnipeg. The additive factor of smoking did not overshadow the effect of pollution in this Tale of Two Cities. Both of these factors appear to be operative at a high level of intensity, and this autopsy analysis implies a dose response relationship to air quality. Tobacco smoke may have a cumulative

TABLE 2 -- AIR POLLUTION EMISSIONS

(thousand tons/year)

	SULFUR OXIDES	NITROGEN OXIDES	HYDRO- CARBONS	PARTI- CULATES
ST. LOUIS	455	138	374	147
WINNIPEG	36	20	62	82

TABLE 3 -- METEOROLOGICAL CHARACTERISTICS

	DEGREE DAYS BY YEARS	PRECIPITATION (INCHES)	WIND SPEED (MPH)
ST. LOUIS	4699	36.8	9.3
WINNIPEG	1679	20.4	12.7

or synergistic action, acting as a non-specific irritant, paralyzing normal defence mechanisms and allowing the accumulation of specifically harmful atmospheric pollutants.

Resting on morphologic evidence for the establishment of the diagnosis of emphysema, and graded by comparable mensural methods applicable to whole lungs, recent evidence suggests that there are significant geographical differences in the prevalence of anatomic emphysema. London apparently has a high incidence of emphysema; the same observer records slightly less in Edinburgh (37). Cardiff is in the same band as Edinburgh and Tokyo (38), Boston and St. Louis (9) appear to be in the same range as Cardiff and Edinburgh. Erlangen, Germany (39), a non-industrial community, has much less anatomic emphysema than Cardiff. Heard, who examined cases of emphysema in 100 consecutive autopsies in both London and Edinburgh, concurred that in Central Africa, specifically Nigeria, where he repeated the study, emphysema was uncommon and limited in extent (40). Most of these investigated cases were in smokers, but the differences in extent and severity of the anatomic emphysema revealed by these studies suggest the presence of an urban factor, probably pollution. Bronchial metaplasia has long been held to be pre-eminently a lesion of the smoker. Current studies, revealing a high incidence of this cytological response in 50% of non-smoking, middle aged, males living in New York city, infer an association with air pollutants (41). Unfortunately, these are incomplete comparisons, because the various factors that can influence a statistical analysis are difficult to control in such diversified studies. Although the many variables that influence the life style of the city dweller emphasize the basic difficulties inherent in geographic field studies, an international survey, obedient to the demands of the statistician, might close some of the hiatuses in our knowledge on air pollution and its relationship to chronic pulmonary disease. It would be interesting, for instance, to know the prevalence of chronic bronchitis in Istanbul, where most of the air pollution essentials are present. The visibility index is sharply reduced, partly due to the Bosphorus ferries burning soft coal, with additive contributions through improper carburetion of gasoline from the innumerable old automobiles on the crowded narrow streets, and the photochemical possibilities from the juxtaposed Sea of Marmara and the many months of warm temperatures.

It is the practice in certain large industrial cities (e.g., Toronto) to publish todays' pollution index in the daily newspaper:

"Sulphur dioxide air pollution average 11.2 parts per hundred million during the past 24 hours.

"Acceptable: Zero to 25 parts. Unacceptable: 26 to 39 parts. Dangerous: Higher than 40 parts."

But the overriding question is whether levels of "zero to 25 parts per 100 million" can be called acceptable without knowing the cumulative effects of chronic exposure to these low doses upon biologic tissues.

Neither of the two main generalized patterns of emphysema has a greater degree of association with the community air pollution level. While generalized centrilobular emphysema is usually associated with bronchiolar damage and thought to be dependent to some extent upon the degree of community pollution, and the increased pigmentation around these centrally located bronchioles may be a marker of pollution, the causal relationship of pigment to the damaged bronchiole has as yet to be established. X-ray diffraction analysis of the pigmented areas in centrilobular emphysema has revealed a number of elements found in air pollution, including titanium, aluminum, lead, tungsten, chromium, and strontium (9).

JOHN P. WYATT

CHRONIC LUNG DISEASE IN CHILDREN

There is good evidence from pathological and statistical studies that 'conditioning' or 'imprintment' factors during early life play a role in subsequent adult respiratory illnesses. Nutritional and, perhaps, environmental experiences of a person during his early formative years may influence respiration ailments throughout that individual's life (42). According to the recent studies carried out in various parts of the world, children (before smoking age) living in more highly polluted areas show decreased pulmonary function when compared to the children living in areas with less pollution. A survey of respiratory diseases in over 10,000 children, aged 6 to 10 years, in England and Wales, has shown that the higher the local level of air pollution, and the lower the socio-economic status, the greater the incidence of chronic bronchitis. Sherwin (43) in California contends from autopsy studies that "bronchiolar" lesions arise in the first decade of life.

Follow up studies of workers exposed to the fumes of explosive shells reveal a high incidence of severe disabling emphysema (44), as do a high proportion of coal miners exposed to nitrous fumes from underground shot firing. There appears to be an augmented incidence of panlobular emphysema in cadmium workers (45). Blackburn's studies (46) on New Guinea natives related their bronchitis and emphysema to the irritating effects of aldehydes produced by wood burning in their unventilated thatched huts. The pattern of emphysema seen in illustration 5A in their article is suggestive of infectional bronchiolitis and centrilobular emphysema. The extensive degree of fibrosis and the implication of its onset in childhood, as well as the excessive carbon pigmentation, suggested a long term relationship between the lung alterations and this particular form of air pollution.

Experimental Evidence

The dissolutive portrait of emphysema has not been experimentally produced with any of the common pollutants in our environment. With respect to experimental models, apart from the unnatural bludgeon of the intratracheally instilled vegetable protease, papain, producing a generalized emphysema (47), all of the allegedly emphysema animal models, utilizing chlorine, phosgene, nitrogen dioxide, cadmium, etc., have been hedged by terminology such as "...emphysema-like or pseudoemphysema". Although these studies fell short of producing a uniform lung-destroying effect, there was evidence of lung irritation in all of them.

REFERENCES

1. Evelyn, John. (1772). Fumifugium or the inconvenience of the aer and smoake of London, printed by W. Godbid for Gabriel Bedel and Thomas Collins, London, 1661; Reprinted for B. White at Horace's Head in Fleet Street.

2. Firket, J. (1931). Bull. Acad. Roy. Med. Belg. 11: 683. Cited from: A. C. Stern (ed.). Air Pollution I, (2nd ed.), Academic Press, New York, 1968.

3. Schrenk, H. H. (1949). Air pollution in Donora, Pennsylvania; Epidemiology of the unusual smog episode of October, 1948. P.H.S. Division of Industrial Hygiene, Public Health Service Bulletin No. 306, U. S. Govt. Printing Office, Washington, D. C.

4. Ministry of Health. (1954). Mortality and morbidity during the London fog of December 1952. Reports on Public Health and Related Subjects No. 95, H. M. Stationery Office, London.

5. Greenwald, I. (1954). Effects of inhalation of low concentrations of sulphur dioxide on man and other animals. Arch. Ind. Hyg. Occ. Med. 10: 455.

6. Frank, N. R. and Speizer, F. E. (1965). SO_2 effects on the respiratory system in dogs. Arch. Env. Health 11: 624.

7. Balchum, O. J., Dybicki, J. and McNeely, G. R. (1960). The dynamics of SO_2 inhalation. Arch. Ind. Health 21: 564.

8. Lamb, D. and Reid, L. (1968). Mitotic rates, goblet cell increase and histochemical changes in mucus in rat bronchial epithelium during exposure to sulphur dioxide. J. Path. Bact. 96: 97.

9. Wyatt, J. P., Fisher, V. W. and Sweet, H. C. (1964). The pathomorphology of the emphysema complex. Amer. Rev. Resp. Dis. 89: 533.

10. Wagner, W. D. et al. (1965). Experimental study of the threshold limit of NO_2. Arch. Env. Health 10: 455.

11. Blair, W. H., Henry, M.C. and Ehrlich, R. (1969). Chronic toxicity of nitrogen dioxide. II. Effect on histopathology of lung tissue. Arch. Env. Health 18: 186.

12. Kleinerman, J. and Wright, G. W. (1961). The reparative capacity of animals' lungs after exposure to various single and multiple doses of nitrite. Amer. Rev. Resp. Dis. 83: 423.

13. Haydon, G. B. et al. (1967). Nitrogen dioxide-induced emphysema in rabbits. Amer. Rev. Resp. Dis. 95: 797.

14. Jaffe, L. S. (1967). The biological effects of ozone in man and animals. Am. Ind. Hyg. Assoc. J. 28: 267.

15. Scheel, L. D. et al. (1959). Physiologic, biochemical, immunologic, and pathologic changes following ozone exposure. J. Appl. Physiol. 14: 67.

16. Gardner, D. B. et al. (1970). Pulmonary changes in 7,000 mice following prolonged exposure to ambient and filtered Los Angeles air. Arch. Env. Health 20: 310.

17. Bowden, D. H. and Wyatt, J. P. (1970). Lung injury and repair: A contemporary view. In: S. C. Sommers, (ed.). Pathology Annual, Appleton-Century-Crofts, New Jersey.

18. Wyatt, J. P. (1953). Non-silica pneumoconiosis. In: J. F. McManus, (ed.). Fundamentals in Medical Progress, Lea and Febiger, 279.

19. Staub, N. C. (1970). The pathophysiology of pulmonary edema. Human Path. 1: 419.

20. Pepys, J. (1969). Hypersensitivity diseases of the lungs due to fungi and organic dusts. In: S. Karger, (Pub.). Monographs in Allergy, Basel, Switzerland.

21. Scadding, J. G. (1970). Lung biopsy in the diagnosis of diffuse lung disease. Brit. Med. J. 2: 557.

22. Elwood, P. C. (1965). Respiratory symptoms in men who had previously worked in a flax mill in Northern Ireland. Brit. J. Ind. Med. 22: 38.

23. El Batawi, M. A. et al. (1964). Byssinosis in the Egyptian cotton industry: changes in ventilatory capacity during the day. Brit. J. Ind. Med. 21: 13.

24. Lawther, P. J. and Waller, R. E. (1970). Air pollution and exacerbations of bronchitis. Thorax 25: 525.

25. Lambert, P. M. and Reid, D. D. (1970). Smoking, air pollution, and bronchitis in Britain. Lancet I: 853.

26. Fattal, G. A. and Wyatt, J. P. (1969). Intensive Care Unit and the pathology of progress. In: S. C. Sommers, (ed.). Pathology Annual, Appleton-Century-Crofts, New York, 43.

27. Christensen, O. W. and Wood, C. H. (1958). Bronchitis mortality rates in England and Wales and Denmark. Brit. Med. J. I: 620.

28. Mork, T. (1962). A comparative study of respiratory disease in England, Wales, and Norway. Acta Med. Scand. 172: (Suppl. 384).

29. Fairbairn, A. S. and Reid, D. (1958). Air pollution and other local factors in respiratory disease. Brit. J. Prev. Social Med. 12: 94.

30. Zeidberg, L. D., Horton, R. J. M. and Landan, E. (1967). The Nashville air pollution study. Arch. Env. Health 15: 214.

31. Toyama, T. (1964). Air pollution and its health effects in Japan. Arch. Env. Health 8: 153.

32. Lave, L. B. and Seskin, E. P. (1970). Air pollution and human health. Science 169: 723.

33. Kaspar, J. and Tichy, L. (1966). International symposium on control and utilization of sulphur dioxide and fly ash from the flue gases of large thermal power plants, Czechoslovakia, 1965. Proc. State Commission for Technology, Praha, 399. (Cited in A. C. Stern, (ed.), (1968). Air Pollution vol. I, Academic Press, New York, 612).

34. Ferris, B. G. and Anderson, D. O. (1962). The prevalence of chronic respiratory disease in a New Hampshire town. Amer. Rev. Resp. Dis. 86: 165.

35. Anderson, D. O., Ferris, B. G. and Zickmantel, R. (1965). The Chilliwack respiratory survey, 1963. The prevalance of respiratory disease in a rural Canadian town. Can. Med. Assoc. J. 92: 1007.

36. Ishikawa, S. et al. (1969). The "emphysema profile" in two midwestern cities in North America. Arch. Env. Health 18: 660.

37. Heard, B. E. (1970). Emphysema in Edinburgh. Pathologia et Microbiologia, and Supplementum 35: 167.

38. Yamanaka, A. (1970). Pulmonary emphysema in Japan. Pathologia et Microbiologia, and Supplementum 35: 161.

39. Heard, B. E. (1969). Pathology of Chronic Bronchitis and Emphysema. Churchill, London.

40. Alli, A. F. (1970). Pulmonary emphysema in Ibadan, Nigeria -- A preliminary report on a pathological study of 114 unselected necropsies. Pathologia et Microbiologia, and Supplementum. 35: 170.

41. Spain, D. M. et al. (1970). Metaplasia of bronchial epithelium. J. Amer. Med. Assoc. 211: 1331.

42. Reid, D. D. (1970). Beginnings of bronchitis. Brit. Med. J. II: 190.

43. Sherwin, R. P. (1970). Emphysema: The "hole" truth. Arch. Env. Health 21: 699

44. Kennedy, M. C. (1970). Emphysema in coalworkers. Brit. Med. J. IV: 115.

45. Gough, J. (1960). Emphysema in relation to occupation. Indust. Med. Surg. 29: 283.

46. Woolcock, A. J. et al. (1970). Studies of chronic (non-tuberculous) lung disease in New Guinea populations: The nature of the diseases. Amer. Rev. Resp. Dis. 102: 575.

47. Gross, P. et al. (1968). Experimental emphysema -- Effect of chronic nitrogen dioxide exposure and papain on normal and pneumoconiotic lungs. Arch. Env. Health 16: 51.

CHAPTER 10. HOST VARIABLES IN PULMONARY RESPONSES TO THE ENVIRONMENT

GARETH M. GREEN, University of Vermont,
College of Medicine, Burlington, Vermont

A prominent shortcoming of much environmental toxicology is the view that these agents are single or primary etiologic factors in the development of chronic pulmonary disease. A major characteristic of nonspecific pulmonary disease is that multiple etiologic factors are identifiable by epidemiologic study. These factors, including environmental agents, operate simultaneously or sequentially in combinations to produce the pulmonary disease. Environmental factors, moreover, are by no means solely determinative in the expression of the disease. Perhaps of equal, or even overriding, etiologic importance in the determination of the pathogenicity of inhaled and ingested materials from the environment, are expressions of constitutional factors in the exposed person. Constitutional factors appear to be molded in part by heredity, as expressed through biochemical and enzyme function, cell metabolism, and immunologic capability of both cellular and humoral components. Life experiences, including childhood and adult illnesses in the lungs or other organs, socioeconomic conditions, and habits also modify host responses to environmental factors and thereby alter their expression in respiratory disease.

Experimental studies of environmental toxicology ought to follow more closely the investigational methodology of epidemiologists, who have long ago learned to take into account the interactions of multiple variables in the production of a final effect. By contrast, current studies of environmental toxicology are still largely concerned with the specific effects of single chemicals, on single host tissues or functions. This chapter will attempt to show why it is time to move toward the development of experimental systems and methods which will include the analysis of interacting variables in the determination of the biologic and health effects of environmental agents.

It is difficult to comprehend mechanisms of pathogenesis of disease which involve multiple variables without some stable unifying hypothesis. The concept of host defense mechanisms in the lung provides such a stable, unifying, mechanistic link between the expression of constitutional factors attributable to a specific and identifiable organ function, and the toxicologic effects of environmental agents. The concept provides a framework for expressing both the direct effects of environmental agents on lung tissues, and also the indirect effects of environmental factors such as dietary ingestants, surface exposures, physical variables, etc., on lung tissues through effects on other body systems. (See Chapter 7.) Interactions among multiple host conditions and environmental agents can be defined in functional terms by quantitative analysis of the several cell and tissue mechanisms which comprise lung defenses. Interactions among the several cell mechanisms of defense can then be examined in similar fashion. In this way, the significance of endogenous and environmental interactions can be quantitatively defined through significant functional mechanisms in the host.

DEFENSE MECHANISMS

In order to pursue this argument in specific terms, the defense mechanisms under consideration will be briefly reviewed (1). The three classes of defense mechanism in the lung include: (a) aerodynamic filtration, (b) transport, and (c) in situ detoxification.

AERODYNAMIC FILTRATION -- serves to protect the alveolar membrane from the environment by causing the precipitation and absorption of environmental gases and particles on relatively impervious and "tough" bronchial membranes. The components of aerodynamic filtration include: (a) inertial impaction, due to turbulence and changes in direction of particle flow; (b) sedimentation, due to gravitational action on particles suspended in non-moving portions of the airstream; (c) Brownian movements; and (d) gaseous absorption by airways proximal to the respiratory membrane. (See Chapters 3 and 6.)

TRANSPORT MECHANISMS -- serve to carry particles out of the lung and comprise the expiratory airstream itself, the mucociliary apparatus, and probably three component mechanisms in the alveoli involving surface fluid transport, interstitial fluid movement to the bronchial tree, and movement of particles to perivascular positions. (See Chapter 4.)

IN SITU DETOXIFICATION MECHANISMS -- include surface effects of substances in lung lining layer on particle surfaces, phagocytosis, and intracellular detoxification. Surface effects include wetting effects, surface activity by alveolar lining fluid lipids, specific and nonspecific opsonic effects, and the action of surface enzymes such as lysozyme. Phagocytosis is divisible into uptake and intracellular digestive activity, because this is clearly a two-step process which involves distinct metabolic pathways and separable physical-chemical events. Genetic defects may influence one pathway without influence on a second.

In summation, these systems could be referred to as clearance mechanisms, in a broad functional sense of the term, since they involve the removal of physical material from the inhaled airstream and subsequently from the surface of the broncho-alveolar surface, and also the removal of toxicity by means of in situ detoxification activity. It is obvious, however, that clearance is not a single physiologic function, and that each of the component parts may be deranged by itself or in combination with one or more of the other components. For example, defects may be localized to single enzymatic deficiencies in the hexose monophosphate shunt preventing the bactericidal action of a cell (2). On the other hand, infective agents such as viruses may alter the aerodynamic filtration of inhaled particles by an effect on respiratory pattern, the transport of particles by destroying the mucociliary epithelium, or phagocytosis by direct or indirect action on alveolar macrophages. Host-environment interactions may be examined by looking at the multiple effects of the single environmental agent on many host systems, by examining the effects of multiple environmental agents on a single host system, and by a combination of both methods. It will help to begin with a one-to-one analysis and then build on that base to consider multiple host and environmental factor interactions.

METHOD OF MEASUREMENT

The method of aerosol challenge by radiotracer labelled bacteria permits the simultaneous study of: (a) aerodynamic filtration through examination of deposition patterns; (b) transport through serial analysis of tracer decline in the lung; and (c) in situ detoxification through study of the alterations in viability of bacteria within the lung. The tracer serves not only as a measure of transport but also as a physical marker of the quantity of bacterial matter living or dead in the lung at any time. Quantitative study of bacterial viability, therefore, when related to the tracer count, provides an index of the bactericidal or detoxification activity of the lung. This bactericidal activity is attributable largely to the phagocytic activity of alveolar macrophages as a first line of defense, and to other cells such as polymorphonuclear leukocytes as a secondary mechanism. The tracer ratio method is summarized very simply in the accompanying formula (Table 1). Subsequent discussion will be concerned largely with endogenous alterations of the several components of the lung defense system, to demonstrate the wide variation that may be induced in the systems even without the addition of environmental chemicals. It is on this variable host background that the toxicology of environmental agents must be studied.

BASELINE MEASUREMENTS IN NORMALS

Studies in healthy animals have shown that there is a remarkable uniformity in the function of the several components of the lung defense mechanisms, including deposition patterns, transport rates, and phagocytic activity (3,4). Defensive activity

TABLE 1 -- CALCULATION OF BACTERICIDAL DETERMINATION
IN EXPERIMENTAL ANIMALS
____

Pulmonary bactericidal value =

$$100 \left(1 - \frac{\text{bact. count (animal)} \times \text{radiotracer count (aerosol)}}{\text{bact. count (aerosol)} \times \text{radiotracer count (animal)}} \right)$$

follows a normal distribution in such sample populations and it is possible to define normality and abnormality in statistical terms. Different colonies of animals show similar distribution patterns, although absolute mean levels will vary depending on: (a) the genetic characteristics of the responsible cell systems as expressed through metabolic and enzymatic activity; (b) the immunologic experience of the cell and systems and, thus, the host; and (c) the physical, chemical, and metabolic characteristics of the inhaled particles. Population differences in immunity may result from differences in experience with specific microorganisms between different populations, or differences in less specific exposures such as differences in gut flora (5). Similarly, differences may be observed in animal groups from different breeders and in barrier sustained vs. standard animal colonies. Systematic evaluation of the influences of such differences on the effects of environmental agents on host defense mechanisms has not been made.

EFFECTS OF SINGLE AGENTS

Much evidence has accumulated from many sources (6-10) regarding the effects of many different agents on the functional activity of: (1) overall resistance to infection; (2) mucociliary activity; and (3) phagocytosis and bactericidal activity in the lung. Thus, inhaled gases such as ozone or nitrogen dioxide; environmental alterations such as hypoxia or high altitude; drugs such as ethanol or immunosuppressive agents and corticosteroids; inflammatory diseases such as that caused by viral infection; metabolic disease produced by renal failure or hydrocarbon aspiration; physical effects such as dehydration; inhalation of complex mixtures such as tobacco smoke; and many other defined exposures and conditions enhance mortality from infection, and susceptibility to infection, presumably through depressant activity on transport, phagocytosis, or humoral immune mechanisms. It is not the purpose of this chapter to review all this information. Rather, four conditions will be mentioned to illustrate mechanisms. These four will be ethanol, hypoxia, acute virus infection, and metabolic disease produced by acute renal failure.

Ethanol demonstrates the dose response relationship between the toxic agent and the bactericidal function of the lung (11). Hypoxia suppresses lung bactericidal activity through its effects on cell metabolism. Its effect can be reproduced in vitro, by reduction of oxygen tensions in cell cultures of alveolar macrophage cells. Alveolar macrophages, in contrast to other phagocytic systems in the host, are highly dependent on oxidative metabolism, and are relatively susceptible to reductions of oxygen tension.

Virus infection shows a unique time relationship in its effects on bactericidal mechanisms in the lung (12,13). Thus, in experimental influenza A viral infection in mice, bactericidal activity is maintained during the first week of viral infection when viral replication is at its peak. On the other hand, at approximately the 6th-7th day there is a sudden inhibition of bactericidal activity in the lung, induced by the virus infection. This inhibition persists for 2-3 days and then the bactericidal activity suddenly returns to normal levels. These findings in virus infection demonstrate the role of viable organic agents and the specificity of mechanism implied by the timing of the inhibitory event.

Finally, the inhibition of bactericidal activity by nephrectomy (14) is of interest because: (a) it involves metabolic disease, namely acidosis (15); (b) it has been clearly demonstrated that the effect is limited to phagocytic activity and not to transport; and (c) these experiments demonstrate that suppression of the bactericidal activity alone is sufficient to account for the emergence of bacterial multiplication, even when transport mechanisms in the lung are preserved (16). The study of nephrectomy also demonstrates that the endogenous disease state of the host must be defined and quantified in any study of environmental effects on the lung. Thus, the presence of an underlying disease may alter the host response to a given environmental agent. This consideration may be particularly relevant for the effects of environmental factors in the expression or complication of chronic obstructive pulmonary disease.

INTERACTIONS AND MULTIPLE EFFECTS

The role of immunity in mitigating the effects of single agents on the lung is illustrated by the demonstration that the effects of ethanol may be reversed in specifically-immunized animals. Thus, in addition to the interaction of endogenous host disease and environmental factors, interactions also can be demonstrated which depend on the prior experience of the host, as well as on current exogenous toxic exposure.

A given environmental alteration may produce different effects under different conditions. For example, hypoxia, which inhibits the bactericidal action of alveolar macrophages against staphylococcal species, has no inhibitory effect on the phagocytosis of proteus organisms, at least in non-immune animals (17). Thus, exposure to hypoxia would be expected to have a greater effect on the staphylococcal than on the proteus population of the lung. It appears, from a comparison of the bactericidal action against staphylococcus vs. proteus, that there is an oxygen-independent component to the lung's in situ detoxification mechanism. This adds one more variable to the multiple components of pulmonary defense mechanisms that must be considered in an analysis of environmental effects on the lung.

Finally, a series of experiments should be mentioned which utilized laboratory rats with or without exposure to silica, and subsequent challenge with bacterial aerosols. They demonstrate the interacting effects of multiple variables on the defensive mechanisms of the respiratory tract. The presence of chronic bronchial or pulmonary parenchymal disease, which is very common in colonies of laboratory animals, is thought to represent an important variable in the response of animals to challenge with infective or inorganic aerosols. For example, exposure of standard rats to sulfur dioxide and other inhaled agents is reported to be complicated by a high incidence of pneumonia in the lung (18). This observation suggests that such gases are highly effective in suppressing host defense mechanisms in animals with preexisting chronic respiratory disease and chronic respiratory infection. Because of this behavioral and histopathologic difference between standard and specific pathogen free (SPF) rats, the two populations were exposed to clouds of silica dust (19). Rather surprisingly, no difference was found in dust retention between the two strains of animals; if anything, the standard rats with chronic respiratory disease were suspected of showing slightly more active excretion of the inhaled dust. When, however, the same animals were tested for their phagocytic activity by the bactericidal method, it was found that the two undusted populations showed significant differences, in that approximately 25% of the standard rats showed abnormally retarded bactericidal activity in the lung.

On the other hand, of the two silica dusted groups, the standard laboratory rats actually showed an improvement in their bactericidal activity. This rather surprising finding indicates that even a fibrogenic particle may not necessarily be harmful to all cellular functions of the lung. More extensive studies of experimental silicosis have supported these findings, in that animals with extensive experimentally-induced silicosis showed little or no depression of phagocytic activity in the lung (20).

Of related interest to this series of experiments are the studies of Ferin (21), who demonstrated in rats that the inhalation of tobacco smoke impaired the clearance of retained material (silica) in the alveoli, and suggested that this inhibition was due to suppression of the phagocytic activity of alveolar macrophages. His data suggest that tobacco smoke predisposes to damage by silica, reduces the phagocytic uptake of the particles, and thus permits the particles to induce their fibrogenic response by avoiding the detoxification mechanism of the macrophage. At any rate, other studies have shown that tobacco smoke does have a direct inhibiting (7) effect on pulmonary clearance in vivo and on alveolar macrophage phagocytosis in vitro (22). Thus, in the study of human disease, background information such as the existence of chronic disease, cigarette smoking habit, virus infection, etc., must be considered in the analysis of effects of environmental agents on the respiratory tract.

Another source of multiple variation of effect comes from the single or multiple nature of the inhaled particulate. When two species of bacteria are inhaled into the lung, the detoxification mechanisms operate independently on each species, handling each as an independent load. Similarly, the introduction of a single exogenous agent, such as ethanol, in the presence of a mixed particle load results in the simultaneous expression of the effects of the exogenous agent on the handling of the individual particles. In the experiment referred to above, staphylococci and proteus organisms in the mixed inoculum are handled as they would be if present alone. Similarly, when ethanol is introduced into the animals, detoxification of the staphylococcal organism, though inhibited, nevertheless proceeds; while detoxification of the proteus organism is completely inhibited. Therefore, when the two organisms are present and a single agent is introduced from the outside, the proteus organism multiplies and emerges as the single pathogenic agent. The lung thus possesses its own self-selecting mechanism for emergence of a single inhaled particle as pathogen from a mixed inoculum. That the interaction of immunity to the proteus organism can protect the host against effects of ethanol, even in mixed infection, is inferred from these data and from previous data showing the effects of immunity in preventing the adverse effects of ethanol intoxication.

The data thus indicate that the outcome of a single encounter, or of a continuing encounter, between host and environment is dependent on multiple interacting variables within the host and within the environment. In the case of an infectious disease, the outcome is determined by: constitutional factors inherent in cell physiology; the presence or absence of preexisting acute or chronic disease in the lungs or other organ systems; the presence or absence of a simultaneously acting exogenous or endogenous toxic environmental substance such as ethanol, tobacco smoke, air pollutant, etc.; and, finally, the nature and specific characteristics of inhaled particles which challenge the lung at any particular time. These considerations help to explain the mechanism and etiologic role of multiple host and environmental factors in the pathogenic expression of acute and chronic pulmonary disease. Thus, the toxicity of single agents, whether from the environment or originating within the host, may be entirely different in the compromised host from that in the healthy person. It is difficult, therefore, to speak of single pathogenic and etiologic agents in the development of chronic disease in the lung. One must think, instead, in terms of multiple, simultaneously, and sequentially interacting variables, which may produce effects when acting in concert although virtually harmless when acting as single agents. Determination of toxicity of environmental variables, therefore, requires the use not only of simple single toxicologic models, but also models in which multiple interactions are an intrinsic part. Indeed, the effects of environmental agents of diseased animals might be more pertinent for much of the problem of human responses to the environment than reliance solely on cytotoxic effects or single physiologic parameters in healthy animals.

REFERENCES

1. Green, G. M. (1970). The J. Burns Amberson Lecture -- In defense of the lung. Amer. Rev. Resp. Dis. 102: 691.

2. Holmes, B., Page, A. R. and Good, R. A. (1967). Studies of the metabolic activity of leukocytes from patients with a genetic abnormality of phagocytic function. J. Clin. Invest. 46: 1422.

3. Green, G. M. and Goldstein, E. (1966). A method for quantitating intrapulmonary bacterial inactivation in individual animals. J. Lab. Clin. Med. 68: 669.

4. Laurenzi, G. A. et al. (1964). A quantitative study of the deposition and clearance of bacteria in the murine lung. J. Clin. Invest. 43: 759.

5. Laurenzi, G. A. et al. (1963). Important determinants in resistance to pulmonary infection. J. Clin. Invest. 42: 949.

6. Green, G. M. (1968). Pulmonary clearance of infectious agents. Ann. Rev. Med. 19: 315.

7. Laurenzi, G. A. et al. (1963). Clearance of bacteria by the lower respiratory tract. Science 142: 1572.

8. Rylander, R. (1969). Alterations of lung defense mechanisms against airborne bacteria. Arch. Env. Health 18: 551.

9. Coffin, D. L. et al. (1968). Influence of ozone on pulmonary cells. Arch. Env. Health 16: 633.

10. Dalhamn, T. (1956). Mucous flow and ciliary activity in the trachea of healthy rats and rats exposed to respiratory irritant gases. Acta Physiol. Scand. 36: 123. (Suppl. 123)

11. Green, G. M. and Kass, E. H. (1964). Factors influencing the clearance of bacteria by the lung. J. Clin. Invest. 43: 769.

12. Green, G. M. (1966). Patterns of bacterial clearance in murine influence. In: G. L. Hobby, (ed.). Antimicrobial Agents and Chemotherapy. American Society for Microbiology, Ann Arbor, Michigan, 26.

13. Klein, J. O. et al. (1969). Effect of intranasal reovirus infection on antibacterial activity of the mouse lung. J. Inf. Dis. 119: 43.

14. Goldstein, E. and Green, G. M. (1966). The effect of acute renal failure on the bacterial clearance mechanisms of the lung. J. Lab. Clin. Med. 68: 531.

15. Goldstein, E., Green, G. M. and Seamans, C. (1970). The effect of acidosis on pulmonary bacterial function. J. Lab. Clin. Med. 76: 912.

16. Goldstein, E. and Green, G. M. (1967). Alteration of the pathogenicity of Pasturella pneumotropica for the murine lung caused by changes in pulmonary antibacterial activity. J. Bact. 93: 1651.

17. Green, G. M. and Kass, E. H. (1965). The influence of bacterial species on pulmonary resistance to infection in mice subjected to hypoxia, cold stress, and ethanolic intoxication. Brit. J. Exper. Path. 46: 360.

18. Lamb, D. and Reid, L. (1968). Mitotic rates, goblet cell increase and histochemical changes in mucus in rat bronchial epithelium during exposure to sulphur dioxide. J. Path, Bact. 96: 97.

19. Morris, T. G. et al. (1967). Comparison of dust retention in specific pathogen free and standard rats. In: C. N. Davies, (ed.). Inhaled Particles and Vapours. Pergamon Press, Oxford, 205.

20. Goldstein, E., Green, G. M. and Seamans, C. (1969). The effect of silicosis on murine pulmonary resistance to experimental aerosol infection. J. Inf. Dis. 120: 210.

21. Ferin, J., Urbankova, G. and Vickova, A. (1965). Pulmonary clearance and the function of macrophages. Arch. Env. Health 10: 790.

22. Green, G. M. (1967). Resistance to bacterial infection in experimental simulated respiratory insufficiency. Clin. Res. 15: 345.

CHAPTER 11. INTERACTION OF INFECTIOUS DISEASE AND AIR POLLUTANTS:
Influence of "Tolerance"

DAVID L. COFFIN, National Environmental Research
Center, Environmental Protection Agency,
Research Triangle Park, North Carolina

It is well known that many if not most disease
states of man, animals, and plants are the result of
several deleterious influences acting simultaneously.
For the establishment of an infection, for instance,
there are required not only the specific
microorganisms but accessory factors which make the
particular individual vulnerable to its invasion.
Such supplementary stresses are frequently termed
predisposing influences. They may consist of many
things which weaken our resistance, such as the
pre-existence of other infection, deficiency of
certain dietary elements, the presence of toxicants,
and many other conditions peculiar to a particular
host or a specific environment. The operation of
such factors in infectious disease has been termed
the principle of multiple causality, which has been
discussed in detail by Top (1).

BACTERIAL INFECTIONS

It is now well established that a number of
environmental influences diminish the lungs' ability
to cope with introduced infectious agents.
Manifestations of this phenomenon are lengthening of
the time of residence of viable bacteria in the
lungs, or, in the event that pathogenic bacteria are
employed, increased mortality to infection in
experimental animals. (The basic mechanisms involved
are more fully discussed in Chapter 10).

The role of an air pollutant as a predisposing influence was suggested in 1957 by the observation of increased spontaneous pneumonia in guinea pigs undergoing exposure to ozone (2). Subsequently, other investigators demonstrated in an experimental model that mortality in mice, from both Streptococcus and Klebsiella infections of the lung, could be enhanced when the exposure to the bacterial aerosol was preceded or followed by exposure to ozone. Significant differences in mortality were demonstrated after exposure from 1 to 9 ppm ozone for three hours (3). Mortality was also enhanced in mice by exposure to nitrogen dioxide in a similar model in which mortality enhancement was reported after a three hour exposure to 3.5 ppm for two hours (4). (Table 1).

Experiments conducted to determine the threshold for the ozone effect have indicated that enhancement of mortality from streptococcus infection results from exposure to as little as 0.08 ppm ozone for three hours (5) (Table 2). Enhancement was also obtained following a 4 hour exposure to synthetic auto smog containing 0.15 ppm total oxidant and above (6) (Table 3). Experiments also indicated that superimposition of the stress of reduced environmental temperature, following exposure to the

TABLE 1 -- INFLUENCE OF A SINGLE NITROGEN DIOXIDE EXPOSURE
ON ENHANCING MORTALITY

% Mortality

NO_2 ppm	NO_2 Exposures	Control	% Difference
25	92	40	52*
15	88	33	55*
10	98	48	50*
5	94	45	49*
3.5	98	44	54*
2.5	40	34	6
1.5	57	47	10

* $P = <0.05$

A statistically significant enhancement of mortality from infection via an aerosol of K. pneumoniae exists at 3.5 ppm and above for a two hour exposure (44).

TABLE 2 -- INFLUENCE OF A SINGLE OZONE EXPOSURE ON
ENHANCING MORTALITY

% Mortality

O_3 ppm	O_3 Exposure	Control	% Difference
.76	90	20	70*
.67	90	10	80*
.52	80	13	67*
.43	100	50	50*
.35	78	13	65*
.30	40	3	37*
.29	83	38	45*
.22	65	40	25*
.20	50	8	42*
.18	63	0	63*
.17	45	8	37*
.10	35	8	27*
.08	47	25	22*
.07	20	12	8
.06	30	15	15
.04	15	5	10
.02	37	25	12

* P = <0.05

A statistically significant enhancement of mortality
from infection via an aerosol of streptococcus
pyogenes Group C occurs at 0.08 ppm and above for a
three hour exposure (5).

ozone streptococcus model, further increased the
differences between the ozone treated and control
animals (7).

One of the striking features of experimental
exposure of animals to nitrogen dioxide is the
increased effect with time. When aerosols of
infectious bacteria were applied to animals which had
been exposed continuously or intermittently to NO_2
for three months or longer, mortality differences
were noted at concentrations as low as 0.5 ppm --
approximately a seven fold reduction of threshold
over that seen in the two hour exposure (8) (Table
4). Furthermore, structural changes consisting of
enlarged air spaces were now observed in these
animals (9,10).

TABLE 3 -- INFLUENCE OF ARTIFICIAL SMOG ON ENHANCING MORTALITY

% Mortality

Total Oxidant ppm	Auto Exhaust Exposure	Control	% Difference
.67	25	0	25*
.65	65	25	40*
.63	85	5	80*
.60	55	5	50*
.59	55	5	50*
.53	40	25	15*
.52	60	5	55*
.49	35	20	15*
.48	70	15	55*
.41	53	7	46*
.35	45	10	35*
.29	43	7	36*
.28	60	13	47*
.26	70	10	60*
.16	40	7	33*
.15	37	10	27*
.14	7	0	7
.12	13	7	6
.08	7	4	3

* $P = <0.05$

A statistically significant enhancement of mortality from an aerosol of S. pyogenes Group C exists at oxidant level of 0.15 ppm and above for a four hour period (6).

TABLE 4 -- INFLUENCE OF CONTINUOUS EXPOSURE TO NITROGEN DIOXIDE IN ENHANCING MORTALITY

% Mortality

Time Exposure	NO_2 Exposed	Control	% Difference
9 months	70	54	16*
6 months	88	48	40
3 months	92	64	28*
2 months	78	68	10
1 month	57	43	14
14 days	51	45	6
7 days	68	67	1

* $P = <0.05$

A statistically significant enhancement of mortality from a single aerosol of K. Pneumonia exists for the three month period or longer (44).

VIRUS INFECTION

Increased mortality from influenza infection has been reported in squirrel monkeys when the virus aerosol (A/PR-8) is followed after 24-48 hours by exposure to NO_2 (10-12). This lag is presumably required to permit sufficient time for the establishment of the infection before exposure to the gas, suggesting that the mechanisms involved in the enhancement of viral infection by NO_2 may be different from that involved in bacterial infection. Explanation for such a difference may be given by the results of experiments indicating that exposure to NO_2 reduces the elaboration of interferon, a viral inhibitory substance produced in response to the interaction of tissue cells and virus (13).

In connection with the enhancement of mortality to influenza, and the reduction of interferon production attributable to NO_2 exposure in experimental animals, a recent report of the epidemiological association of influenza morbidity with pollution by NO_2 is of great interest (14).

MODE OF ACTION

The oxidant air pollutants ozone, nitrogen dioxide, and synthetic auto smog probably impose somewhat similar immediate effects in depressing the pulmonary defense to infectious agents. (See Chapter 10.) Since these effects have been most studied for ozone, the subject will be discussed in terms of this gas. Perhaps the most notable feature of exposure to ozone has been the extreme alteration of the dynamics of bacterial decline and gain in the lung. With the Streptococcus Group C used, normally less than 4% of the bacteria initially present still survive after four hours, and the lungs are frequently sterile by ten hours after the cessation of the bacterial aerosol exposure (5). When the mice are pretreated by ozone exposure, this decline in the bacterial population is reduced, and followed later by a growth phase, in which the organisms eventually become more numerous than at the time of initial deposition (5).

A series of experiments performed in the writer's laboratory, comparing bacterial decline, multiplication, invasion of the blood and mortality with ozone concentration, indicated that all were dose related (15). In these experiments controls breathing only ambient air had an average of less than 4% of the initially viable bacteria four hours subsequent to the aerosol exposure, whereas those exposed to 1 ppm or more of ozone had an actual gain at this period. Subsequent to the four hour period there was an acceleration of bacterial growth in the animals receiving the ozone treatment. The phases of bacterial decline, lag, and growth are summarized and related to mortality in Figure 1.

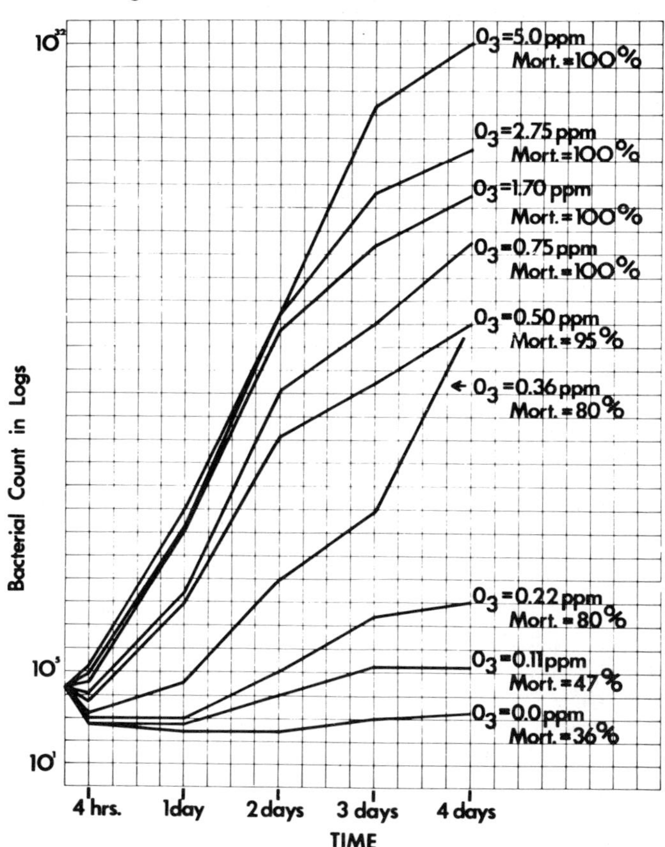

Fig. 1. Influence of ozone exposure on bacterial numbers in the lung. A dose related diminution of bacterial decline, shortening of the lag phase and acceleration of the growth is evident. These changes are proportional to the mortality rate (15).

In these experiments, invasion of the blood by the bacteria was first noted two days after exposure to the bacterial aerosol, and appeared positively related to mortality in both exposure and control groups. Death could be predicted from a positive blood culture (Figure 2). The post exposure time at which median mortality occurred was approximately 4.5 and 7.0 days respectively for the ozone exposed and ambient air controls. Ozone exposure produces slower bacterial elimination, shortened lag phase between bacterial decline and outset of growth, increased growth of bacteria in the lung, and both accelerated and increased incidence of invasion of the blood stream. The earlier deaths in the ozone treated animals appeared consistent with the more rapid bacterial replication in the lung and earlier invasion of the blood (Figure 2).

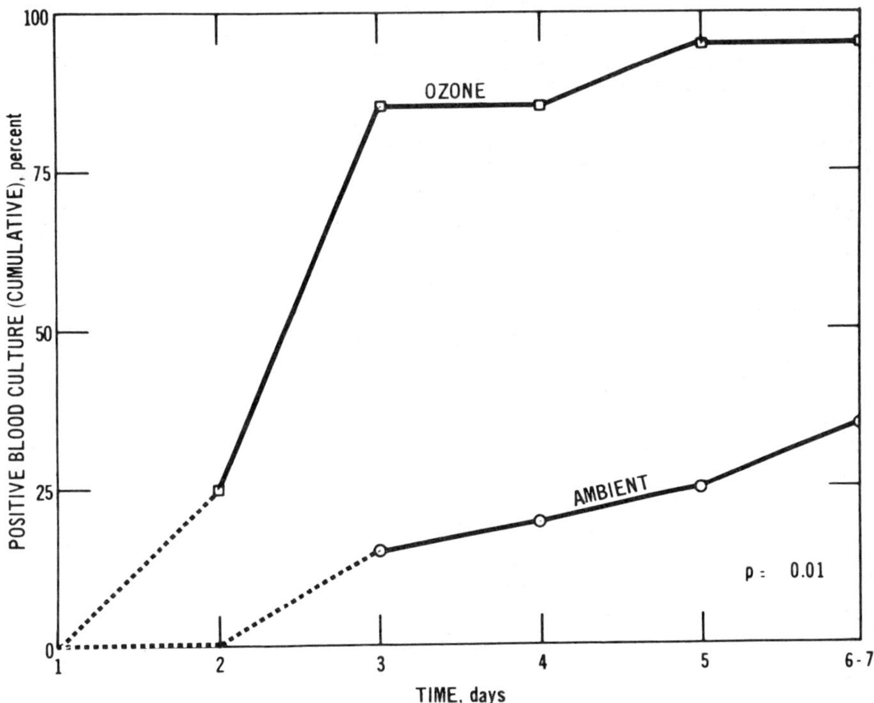

Fig. 2. Influence of ozone exposure on bacterial invasion of the blood. Exposure to ozone accelerates the penetration and augments the number of positive blood cultures. These changes are proportional to the mortality rate.

The action of oxidant air pollutants in predisposing to infection of the lung might be explained in several ways. These include: (1) reduction of the physical removal of the bacterial cells via the so-called mucociliary ladder; (2) the production of edema; (3) alteration of the specific cellular defense mechanism; (4) alteration of the chemical or physical state of the non-cellular milieu; and (5) morphological alterations. (For discussions of the basic mechanism see Chapters 4 and 10.)

It does not appear that depression of the physical removal of bacterial cells plays a major role, since it has been shown that decline in bacterial viability proceeds at a much faster rate than physical removal in the lung (16). While the presence of edematous fluid undoubtedly is important in enhancing bacterial growth, it would seem highly doubtful that appreciable edema of the lungs could have been produced at the low levels of ozone used in these experiments, which are consistent with actual ambient levels in polluted atmospheres. Furthermore, experiments designed to elicit tolerance to ozone indicate that, while tolerance to the formation of edema occurs following a sensitizing dose, no such effect is evident in the specific cellular parameter to be discussed later.

Evidence for specific action of air pollutants on the pulmonary macrophage system is rather strong. This deleterious action on the alveolar macrophage has implications beyond merely reducing the antibacterial defenses, since the cells are usually regarded as the first line of defense against all sorts of material which has gained entrance to the deep lung. They therefore are important for the clearance of toxic particles, carcinogens and the like from the pulmonary alveoli. It has been noted, for instance, that when bacteria are introduced into the lung via an aerosol they are at first primarily free in the alveolar spaces, but subsequently are almost entirely contained within phagocytic alveolar cells (17).

Experiments conducted with air pollutants show that ozone, and to a lesser extent nitrogen dioxide, reduces the number of alveolar macrophages obtainable by pulmonary lavage (18,19) (Figures 3,4). The ability of the macrophage cells to engulf bacteria is also reduced (18) (Figure 5) and their bacteriolytic enzyme activity is lessened (20,21) (Figure 6). Furthermore, ozone exposure appears to exert an influence on a non-cellular component present in the airway, which reduces the ability of alveolar macrophages to survive in artificial media (22). Exposure of rabbits to ozone has been noted to increase the osmotic fragility of alveolar macrophages (23).

ENHANCEMENT BY CHRONIC EXPOSURE

The prolongation of bacterial viability in the lung, and the various deleterious changes in the macrophages conferred by exposure to ozone, and presumably the other oxidant gases, are ephemeral phenomena persisting approximately 24 hours. Such transience would appear natural for all macrophage

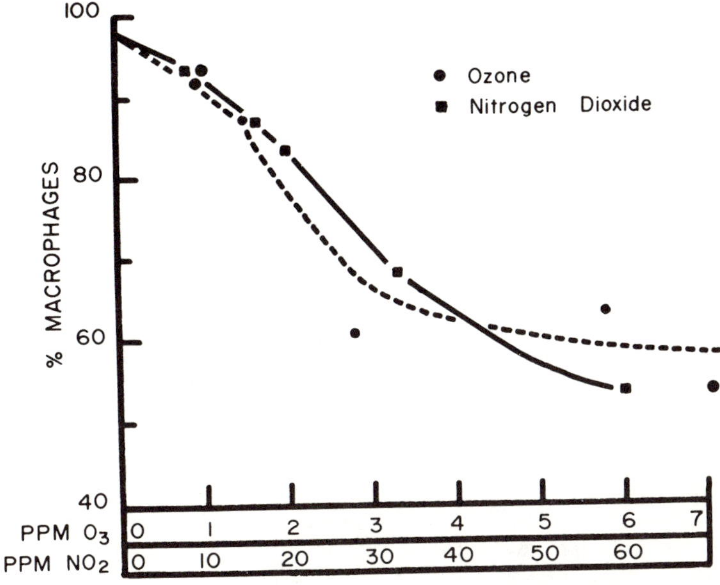

Fig. 3. Influence of exposure to air pollutants on number of pulmonary macrophages of rabbits. Exposure to ozone and nitrogen dioxide reduces the number of pulmonary macrophages obtainable by pulmonary lavage. For ozone (O_3) this effect is noted at 0.9 ppm after a three hour exposure and plateaus above 3 ppm. Approximately ten times the amount of nitrogen dioxide (No_2) is required for an essentially similar effect (18,19).

CHANGES IN CELL POPULATION IN RESPONSE TO OZONE

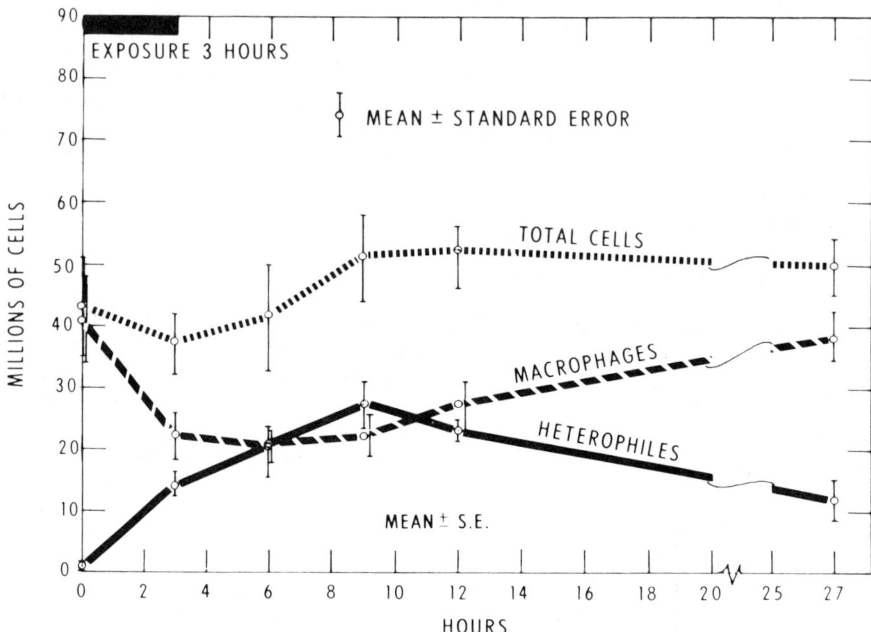

Fig. 4. Persistence of pulmonary cytological alteration induced by ozone exposure. Following exposure there is a reduction of the pulmonary alveolar macrophages and a concomitant increase in polymorphonuclear leukocytes (heterophiles) obtainable by pulmonary lavage. These effects peak at 3-9 hours following exposure, to return to approximately normal values 24 hours after exposure (18).

effects, since the supply in the lung is apparently constantly being replenished by the migration of new cells into the lung, transported via the blood from a bone marrow nidus (24). These new macrophages never having been exposed to the oxidant gas would be expected to react normally.

It would seem, therefore, that some additional injury must be called upon to satisfactorily explain the lowered threshold for enhancement of mortality from infection that is associated with increased exposure time to nitrogen dioxide. It is possible that other cytological changes in the lung may be responsible. Reference has already been made to the enlarged air spaces and other changes noted in mice

DEPRESSION OF PHAGOCYTOSIS BY OZONE

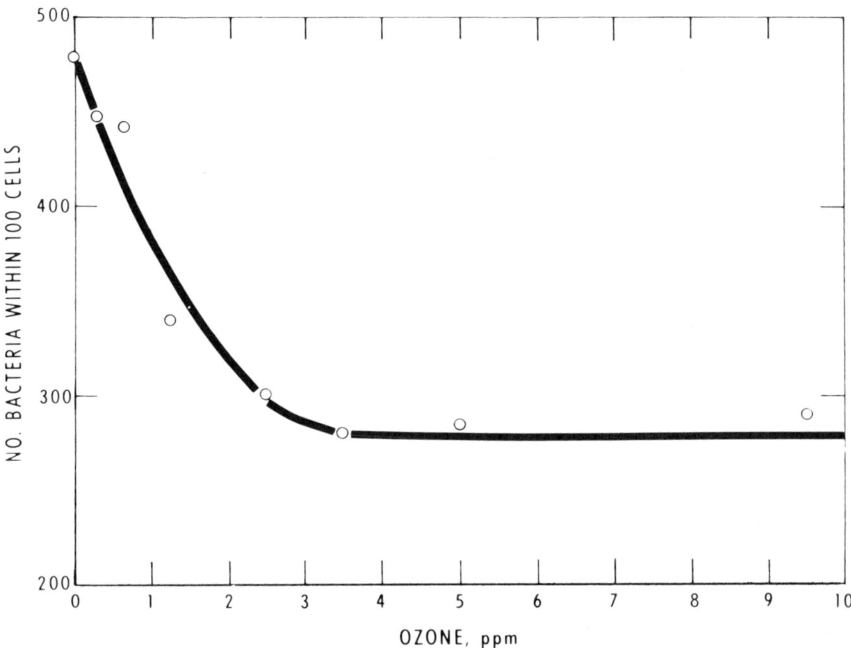

Fig. 5. Influence of ozone exposure on phagocytosis by pulmonary macrophages. A dose related reduction of phagocytosis tends to plateau above 3 ppm (18).

which had undergone chronic exposure to 0.5 ppm NO_2 (9). Continuous exposure of rats to a range of 25-2 ppm nitrogen dioxide causes cellular alteration of the terminal portion of the pulmonary airway. Briefly, these changes consist of loss of cilia from the terminal bronchiolar epithelium, metaplastic thickening of these same cells, and hyperplasia of the lining cells of adjacent pulmonary alveoli. Enlargement of both the total lung and the air space is also attributed to nitrogen dioxide exposure (25,26,27). It seems very probable that such change would lead to increased vulnerability of the centrilobular areas to bacterial invasion, and this might explain the increased mortality following chronic exposure to nitrogen dioxide.

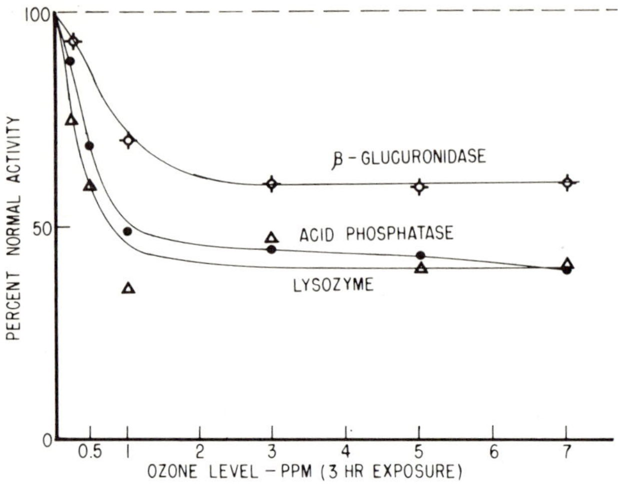

Fig. 6. Influence of ozone exposure on lysosomal enzyme activity. A dose related reduction of activity tends to plateau above 3 ppm (20).

TOLERANCE TO OXIDANT GASES

A phenomenon called "tolerance" has been described for exposure to ozone, nitrogen dioxide, and certain other gases. In the classic experiments, this tolerance or protection develops when a normally lethal dose is preceded by a period of hours or days by a smaller sensitizing or initiating dose (28). Although a great deal of investigation has been carried out on this subject the exact mechanism concerned with its development is still unclear.

Regardless of the mechanism of its initiation, the phenomenon has relevance to air pollution control, since, if it were operant in human beings, it would confer protection against oxidant air pollutants. Most animals tested have yielded evidence of some degree of tolerance development. However, the manifestations of such tolerance have been prevention of mortality or pulmonary edema development initiated by exposure concentrations manifestly greater than those expected in polluted atmospheres. It is of interest that the tolerance development can be induced with an appropriate model so that protection of only one lung occurs (29,30) (Figure 7).

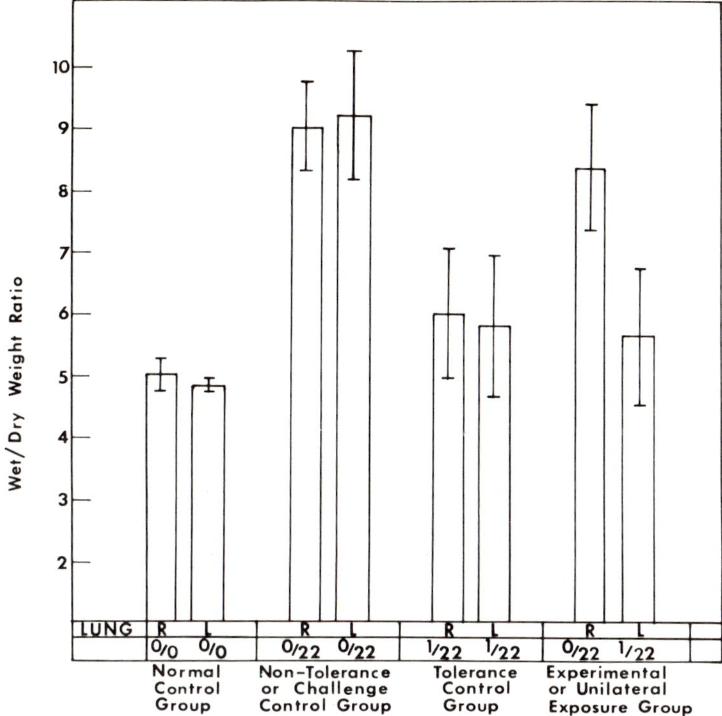

Fig. 7. Influence of "tolerance" on formation of pulmonary edema. Wet weight/dry weight ratios are shown for: (1) normal controls, (2) groups receiving one exposure to 22 ppm of ozone, (3) groups receiving 1 ppm ozone with a second dose of 22 ppm the following day, and (4) groups receiving differential exposure of the right and left lungs by means of catherization in the same sequence as (3). Results shown indicate that tolerance from previous exposure to a low dose blocks edema formation when the whole lung is exposed or when exposure is different unilaterally (30).

When "tolerance" is investigated by means of interaction of pollutants with bacterial infection of the lung, it fails to effectively block the enhancing effect of the gas, even though some protection is evident at the higher levels of exposure studied (5). Furthermore, the effect of tolerance to ozone on the other intrapulmonary defense parameters appears to be non-existent (31). Thus no protection against the influx of polymorphonuclear leukocytes from the blood into the airway, usually considered an index of chemotactic (inflammatory) effect, resulted from "tolerance" (31). Neither did "tolerance" prevent the loss of lysosomal enzyme activity from the alveolar macrophages. These effects are summarized

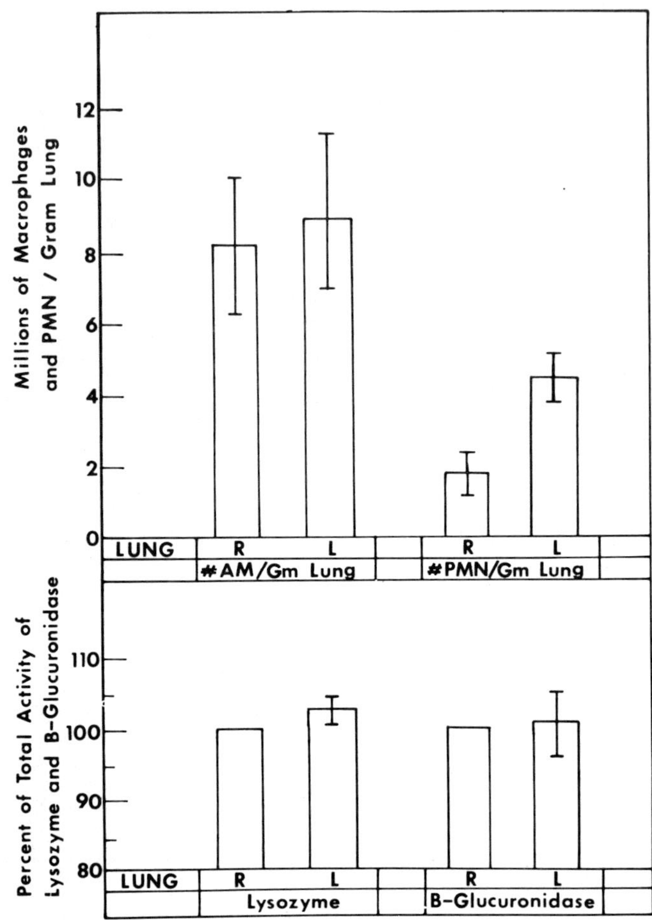

Fig. 8. Influence of "tolerance" on pulmonary cytologic parameters. Differential exposure of the right and left lungs for purposes of assaying "tolerance" shows that this phenomenon has no inhibitory influence on the effect of ozone on number of alveolar macrophages (AM), the polymorphonuclear leukocytes (PMN), and the lysosomal enzymes, lysozyme, and B-glucuronidase (31).

in Figures 7 and 8. The data suggest that the tolerance phenomenon, while effectively preventing the formation of pulmonary edema, is only partially effective against impairment by ozone of the pulmonary defenses against bacterial infection.

INTERACTION OF NONOXIDANT AIR POLLUTANTS WITH INFECTIOUS AGENTS

There is a paucity of information on nonoxidant gases and particles in regard to activation of infection. Despite the apparent association of pulmonary infections in man with air pollutant atmospheres characterized by high concentrations of particles and oxides of sulfur (32,33), few experiments have been reported in the experimental literature.

A series of experiments examined the potential interactions between various dusts and Pneumococcus infection in rats (34). Exposure for 2 to 165 days to smoke and coal dust containing as many as 850 particles per cubic foot, with sulfur content as high as 1.6 ppm, caused no enhancement of mortality from pneumococcus infection. Such findings may suggest that particles of this type are inactive in enhancing infection. In fact, carbon particles are known to increase the number of macrophages in the lung (35). It is conceivable that this would improve the animals' capacity for destroying introduced bacteria rather than diminish it.

Exposure of mice in the writer's laboratory for three hours to sulfuric acid mist with a mass median diameter of less than 1 micron, at concentrations of 80, 150, and 300 mg/m^3, and subsequently to aerosolized virulent streptococci, indicated that only the highest concentration enhanced mortality. Exposure of rabbits to these same concentrations has at no time altered the number of macrophages or polymorphonuclear cells in lavage fluid, even when the animals breathed through fenestrated tracheae. Direct instillation of 0.1 N sulfuric acid into the trachea, however, produced an enormous influx of polymorphonuclear cells. Although the exact reason for the failure of sulfuric acid mist to activate infection is unclear, low penetration into the deeper portion of the lung might be suspected in view of the fact that instillation of acid elicited a marked response. Exposure to sulfur dioxide is known to produce alteration of the pulmonary bronchioles with the production of excess mucus (36). This effect is usually noted only with exposure to high levels, however, and, although the sulfur dioxide effect is potentiated by the addition of particles, the role of this gas in the production of infections of the bronchioles or lung parenchyma is unclear (37). Exposure to sulfur dioxide slows the physical removal of material via the ciliary mucus-flow mechanism (38). A study of the influence of carbon black on removal of bacterial cells from the lung indicated that there was a slight reduction in both physical removal and bacterial kill (39). When both coal dust and sulfur dioxide were administered simultaneously, however, the effect on these two parameters was greater than with either treatment alone (40).

Exposure of rats to lead sesquioxide (10 to 150 $\mu g/m^3$) lowered the number of macrophages obtainable by pulmonary lavage. The effect peaked after eight days of exposure (41).

IMPLICATIONS FOR AIR QUALITY CRITERIA

Frank manifestations of disease are uncommon at levels of toxicant commonly found in natural environments. Experimental work with the infective system, however, indicates that profound alteration of the lungs' defenses against introduced bacteria can be effected by levels of oxidant pollutants well within the range of those occurring in ambient air. Two aspects of the findings of the experimental model are worthy of consideration in relation to the development of air quality criteria information. Firstly, there is the possibility that the model may mimic a like situation which occurs or is likely to occur in man. Necessary components for its duplication spontaneously in man are: exposure to a sufficient concentration of ozone or other oxidant gas, the presence of an anti-infective mechanism capable of being altered by the exposure, and the presence in the environment of infectious agents capable of exploiting the state of diminished resistance (42).

While statistically significant mortality increments are noted in mice at ozone concentrations well below the ambient peaks found in many American cities, it is difficult to relate a given effective concentration from mouse to man. Differences in airway length and complexity, velocity of air movement, the relationship of the volume of air to the area of the target surface in the lungs, and other functional and anatomic species differences would all affect the comparative toxicity of the gas. It would appear that salient information on which to base such an extrapolation could be derived from the construction of appropriate mathematical models. While sufficient functional and anatomic data are present for man, and probably also for the mouse (43), no discussion of such a model has been noted in the available literature.

It is known that the rate of infectivity can be enhanced by oxidant gases in mice (2,4,5,6,8), rats, hamsters (44), and squirrel monkeys (10,11). It thus seems doubtful that man would be incapable of a similar reaction with a like concentration of the gas presented to the target site.

Among the infectious agents eliciting this reaction in animals are Streptococcus pyogenes Group C, Klebsiella pnuemoniae, Diplococcus pneumoniae, and animal adapted influenza A (PR 8). Since the last three are human pathogens, our own species does not appear to lack this component of the model. In essence, the direct application of this model to disease in the human population is still a most crucial matter which must be solved by comparative toxicology, and most importantly, epidemiological data.

Another aspect to be considered in the appraisal of this model is its value simply as an exquisitely sensitive indicator of in vivo biological effect. Increased susceptibility to infection conferred by oxidant gases is a strikingly clear indicator of biological effect at concentrations probably below that demonstrated by any other objective method. The model, therefore, has broader implications than merely showing that susceptibility to infectious disease is increased by oxidants. It indicates that a basic adaptative defensive mechanism of the lung has been impaired. At least a part of this impairment appears as the result of depression of the macrophage system by action of the oxidant. It is generally concluded that the macrophage constitutes one of the most important defensive mechanisms of the deep lung. It has been postulated, for instance, that transport and storage of carcinogen containing particles is carried out through the agency of the pulmonary alveolar macrophage, and their role in the clearance of insoluble particles and destruction or clearance of bacteria is generally accepted. While the bacterial system has been most sensitive in mice, there is reason to believe that the difference in levels of pollutants required to affect other animals (rats, hamsters, squirrel monkeys, and even man) may lie more in lack of virulence of the bacterial invaders selected than in any basic difference in the action of pollutants on the bacterial defense mechanisms. In the writer's laboratory for instance, while mechanistic studies have yielded decrements in many bacterial defensive mechanisms in rabbits, no organisms capable of regularly producing acute lung infection, with the production of subsequent mortality, has been found for this species. Despite a high virulence for mice by other routes, pneumococci have shown a low potential for the production of mortality when the exposure is by means of an aerosol. The difference in threshold for NO_2 in the mouse and hamster from Klebsiella pneumoniae infection, 3.5 ppm and 35 ppm respectively, may be explained by the very low susceptibility of the hamster for this organism (44).

It would indeed appear that differences in infection and subsequent death from combined treatment with various oxidant pollutants and infectious agents, among the various species of animals, hinge more on the virulence and invasiveness of the infectious organism chosen than on the ability of the gas to attack the defense systems of the lung. The sine qua non for the successful operation of this system, as presently used in animals, is the availability of a microorganism which is capable of invading the lung with the production of subsequent mortality. The pollutant then enhances the mortality by depression of the bacterial defense mechanism.

Man is commonly susceptible to several acute bacterial pneumonias, notably from Diplococcus pneumoniae and Klebsiella pneumoniae and influenza virus. These agents have thus far been effective in producing infection in animals in the model under discussion. It would appear reasonable to postulate that exposure to sufficient ozone, oxidant from auto smog, or nitrogen dioxide derived from combustion might predispose human beings to such acute infections of the lung.

It would also appear that the increasing effect of nitrogen dioxide with time has extremely important implications for air quality criteria. The fact that increased susceptibility to acute bacterial infection is associated with subtle anatomical lesions in the terminal airway, raises the point that exposure to oxides of nitrogen may be suspected in the causation of chronic obstructive lung disease in human beings. No definitive data are presently available for areas where association with human disease is best, but low levels of oxides of nitrogen might be predicted to occur wherever there is considerable combustion of fuel (45).

REFERENCES

1. Top, F. H. (1964). Environment in relation to infectious diseases. Arch. Env. Health 9: 699.

2. Stokinger, H. E. (1957). Evaluation of the acute hazards of ozone and oxides of nitrogen. Arch. Ind. Health 15: 181.

3. Miller, S. and Ehrlich, R. (1958). Susceptibility to respiratory infection of animals exposed to ozone. J. Inf. Dis. 103: 145.

4. Purvis, M. R. and Ehrlich, R. (1963). Effect of atmospheric pollutants on susceptibility to respiratory infection. II. Effect of nitrogen dioxide. J. Inf. Dis. 113: 72.

5. Coffin, D. L. et al. (1967). Effect of air pollution on alteration of susceptibility to pulmonary infection. Proceedings of 3rd Annual Conference on Atmospheric Contamination in Confined Space. Wright Patterson Air Force Base, Dayton, Ohio.

6. Coffin, D. L. and Blommer, E. J. (1967). Acute toxicity of irradiated auto exhaust indicated by mortality from streptococcal pneumonia. Arch. Env. Health 15: 36.

7.   Coffin, D. L. and Blommer, E. J. (1965). The influence of cold on mortality from streptococci following ozone exposure. J. Air Poll. Control Assoc. 19: 523.

8.   Ehrlich, R. and Henry, M. C. (1968). Chronic toxicology of nitrogen dioxide. 1. Effects on resistance to bacterial pneumonia. Arch. Env. Health 17: 860.

9.   Blair, W. H., Henry, M. C. and Ehrlich, R. (1969). Chronic toxicity of nitrogen dioxide. II. Effect on the histopathology of the lung tissue. Arch. Env. Health 18: 186.

10. Henry, M. C., Erhlich, R. and Blair, W. H. (1969). Effect of nitrogen dioxide on resistance of squirrel monkeys to Klebsiella pneumoniae infection. Arch. Env. Health 18: 580.

11. Henry, M. C. et al. (1970). Chronic toxicity of NO_2 in squirrel monkeys. Arch. Env. Health 20: 566.

12. Ehrlich, R., Henry, M. C. and Fenters, J. (1970). Influence of nitrogen dioxide on resistance to respiratory infection. In: Inhalation Carcinogenesis Conference, AEC Series 18, U. S. Atomic Energy Commission, Div. Tech. Info.

13. Valand, S. B., Acton, J. D. and Myrvik, Q. N. (1970). Nitrogen dioxide inhibition of viral-induced resistance in alveolar macrophages. Arch. Env. Health 20: 303.

14. Shy, C. M. et al. (1970). Effects of community exposure to nitrogen dioxide. II. Incidence of acute respiratory illness. Air Poll. Control Assoc. J. 20: 582.

15. Coffin, D. L. and Blommer, E. J. (1971). Alteration of the pathogenic role of Streptococci group C in mice conferred by previous exposure to ozone. In: I. H. Silver, (ed.). Proceedings 3rd International Symposium on Aerobiology, Academic Press, New York.

16. Green, G. M. and Kass, E. H. (1964). Factors influencing the clearance of bacteria by the lung. J. Clin. Invest. 43: 769.

17. Green, G. M. and Kass, E. H. (1964). The role of the macrophage in the clearance of bacteria from the lung. J. Exp. Med. 119: 167.

18. Coffin, D. L., Gardner, D. E. and Holzman, R. S. (1968). Influence of ozone on pulmonary cells. Arch. Env. Health 16: 633.

19. Gardner, D. E., Holzman, R. S. and Coffin, D. L. (1969). Effects of nitrogen dioxide on pulmonary cell population. J. Bact. 98: 1041.

20. Hurst, D. J., Gardner, D. E. and Coffin, D. L. (In press). Effect of ozone on acid hydrolases of pulmonary macrophages. J. Retic. Soc. 8: 288.

21. Hurst, D. J. and Coffin, D. L. (1971). Effect of ozone on lysosomal hydrolases in vitro. Arch. Int. Med. 127:

22. Gardner, D. E. et al. (1971). Loss of protective factor for alveolar macrophages when exposed to ozone. Arch. Int. Med. 127: 1078.

23. Dowell, A. R. et al. (1970). Rabbit alveolar macrophage damage caused by in vivo ozone inhalation. Arch. Env. Health 21: 121.

24. Pinkett, M. O., Cowdrey, C. R. and Howell, P. C. (1966). Mixed hematopoietic and pulmonary origin of alveolar macrophages as demonstrated by chromosome markers. Amer. J. Path. 48: 859.

25. Freeman, G. et al. (1968). Lesions of the lung of rats continuously exposed to two parts per million of nitrogen dioxide. Arch. Env. Health 17: 181.

26. Freeman, G. et al. (1968). Pathogenesis of the nitrogen dioxide-induced lesion in the rat lung: A review and presentation of new observations. Amer. Rev. Resp. Dis. 98: 429.

27. Freeman, G. et al. (1968). Environmental factors in emphysema and a model system with NO_2. Yale J. Biol. and Med. 4: 567.

28. Stokinger, H. E. and Coffin, D. L. (1968). Biological effects of air pollution. In: A. C. Stern, (ed.). Air Pollution, 1: 445, Academic Press, New York.

29. Frank, N. R., Brain, M. D. and Sherry, D. E. (1970). Unilateral tolerance and altered elastic behavior in rabbits. Am. Ind. Hyg. Assoc. J. 31: 31.

30. Alpert, S. M. and Lewis, T. R. (1971). Ozone tolerance studies utilizing unilateral lung exposure. J. Appl. Physiol. 31: 243.

31. Alpert, S. M. et al. (In press). Effects of exposure to ozone upon the defensive mechanisms of the lung. J. Appl. Physiol. 31: 247.

32. Douglas, J. W. and Waller, R. E. (1966). Air pollution and respiratory infection in children. Brit. J. Prev. and Soc. Med. 20: 1.

33. Holland, W. W. and Reid, D. D. (1965). The urban factor in chronic bronchitis. Lancet I: 445.

34. Ventinner, F. J. and Baetjer, A. M. (1951). Effects of bituminous coal dust and smoke on the lungs - animal experiments. 1. Effects on susceptibility to pneumonia. Ind. Hyg. and Occup. Med. 4: 206.

35. LaBelle, C. W. (1961). Patterns and mechanisms in the elimination of dust from the lungs. In: C. N. Davies, (ed.). Inhaled Particles and Vapours, 1. Pergamon Press, London, 356.

36. Reid, L.   (1963).   An experimental study of hypersecretion of mucus in the bronchial tree. Brit. J. Exp. Path. 44: 437.

37. Amdur, M. O. (1961). The effect of aerosols on the response to irritant gases. In: C. N. Davies, (ed.). Inhaled Particles and Vapours. 1. Pergamon Press, London, 281.

38. Dalham, T. and Strandberg, L. (1963). Synergism between sulfur dioxide and carbon particles. Studies on absorption and on ciliary movement in the rabbit trachea in vivo. J. Air Water Poll. 7: 517.

39. Rylander, R. (1968). Pulmonary defense mechanisms to airborne bacteria. Acta Physiol. Scand. Supp. 306, 55.

40. Rylander, R. (1970). Studies on lung defenses to infections in inhalation toxicology. Arch. Int. Med. 126: 496.

41. Bingham, E. et al. (1968). Alveolar macrophages: Reduced number in rats after prolonged inhalation of lead sesquioxide, Science 162: 1297.

42. Coffin, D. L. (1970). Study of the mechanisms of the alteration of susceptibility to infection conferred by oxidant air pollutants. In: Inhalation Carcinogenesis AEC Series 18, U. S. Atomic Energy Commission, Div. Tech. Info.

43. Key, C. R. (1966). Morphometry of mouse lungs. Thesis. University of Oklahoma.

44. Ehrlich, R. (1966). Effect of nitrogen dioxide on resistance to respiratory infection. Bact. Rev. 30: 604.

45. Coffin, D. L. (1971). Air pollution: Present and future threat to man and his environment. In: J. M. Pitts and R. L. Metcalf, (eds.). Advances in Environmental Science, II, Wiley Interscience Press, New York.

PART III

SOME SPECIAL PROBLEMS

CHAPTER 12. SMOKING HABITS AND AIR POLLUTION IN RELATION TO LUNG CANCER

E. CUYLER HAMMOND, American Cancer Society
New York, New York

SMOKING

Extensive reviews of the literature (1,2) on the relationship between smoking habits and lung cancer have appeared so recently that there is no point in covering the same ground in detail here. Instead, we will present some previously unpublished data from a large prospective study and briefly summarize the results of other epidemiological, pathological, and experimental investigations.

In 1939 Müller (3) reported that a history of cigarette smoking was far more common in a sample of lung cancer patients than in a sample of patients with other diseases. In the same year, Ochsner and De Bakey (4) observed that nearly all of their lung cancer patients were cigarette smokers. This attracted little attention until the late 1940's when mortality statistics from many countries indicated that death rates from lung cancer had been increasing rapidly during the preceding two or three decades. The concomitant increase in both cigarette smoking and air pollution of certain types suggested that one or the other of these two factors might be the culprit. The association between cigarette consumption and lung cancer death rates in various countries -- and the reportedly higher lung cancer death rates in urban areas than in rural areas -- pointed to these same two factors.

In 1950, Wynder and Graham (5) and Levin et al. (6) reported the results of retrospective studies in which a history of cigarette smoking (particularly heavy cigarette smoking) was found in a far higher proportion of lung cancer patients than control subjects. The same was found in 1952 by Doll and Hill (7) and in many retrospective studies caried out by other investigators in later years.

The results of two prospective studies (8,9), first reported in 1954, confirmed the association between cigarette smoking and lung cancer death rates, as did the results of several later prospective studies (10-14). Findings in all of these studies are in such good agreement that it suffices to present data from just one of them as described below.

Starting on October 1, 1959, volunteer workers of the American Cancer Society enrolled over 1,000,000 men and women and requested each of them to answer a detailed questionnaire including questions on smoking habits, place of residence, occupational exposures, and many other factors. The study area covered 1,121 counties in 25 states. Many of these counties are rural and far removed from any large city; but 16 of the 20 largest cities in the United States, as well as many smaller cities, towns, and suburban areas were included. Nearly 99% of the subjects were traced for the ensuing six years; and at two-year intervals surviving subjects were requested to answer brief questionnaires. Causes of death were ascertained from death certificates. Whenever cancer was mentioned on a death certificate, the doctor, hospital, or cancer registry was requested to supply additional medical information.

Findings on smoking in relation to death rates were last presented after the subjects had been traced for four years (15). The data about to be described cover the entire six-year period and 5,736,868 person-years of exposure to risk (2,472,758 man-years and 3,264,110 woman-years) of subjects aged 35-84 at the start of the study. During the six years, 2,063 of the male subjects and 327 of the female subjects died of lung cancer. The subjects were divided into five-year age groups according to their ages at the time they enrolled in the study.

Death rates were computed for each of these five-year date-of-birth cohorts by dividing the number of deaths during the six years by the person-years of exposure to risk. Age-standardized death rates for broader age groups were computed by averaging the component five-year group rates weighted by the total number of subjects in each component five-year age group.

Findings in Men

Table 1 shows lung cancer death rates of the male subjects classified by type of smoking (lifetime history), and by age at time of enrollment in the study. The death rates were lowest in men who never smoked regularly; somewhat higher in pipe and cigar smokers who had never smoked cigarettes regularly; far higher in cigarette smokers who had also smoked pipes or cigars; and highest in men with a history of smoking only cigarettes.

Mortality ratios were calculated by dividing the death rate of men in each smoking category by the death rate of men who never smoked regularly. For age group 35-84, the mortality ratio was 1.00 for men who had never smoked regularly, 2.23 for men with a history of only pipe smoking, 2.15 for men with a history of only cigar smoking, 8.23 for men with a history of cigarette and other types of smoking, and 10.08 for men with a history of only cigarette smoking.

Table 2 is confined to men who were currently smoking cigarettes regularly at the time that they enrolled in the study (some of them also smoked or had smoked pipes or cigars regularly). They are classified in three different ways: 1) by current number of cigarettes smoked per day; 2) by degree of inhalation of cigarettes smoked; and 3) by age at start of cigarette smoking. These three indices of exposure are highly correlated with each other. For example, men who started cigarette smoking at an early age tend to smoke more cigarettes a day and tend to inhale the smoke more deeply than men who started cigarette smoking later in life (16).

TABLE 1 -- NUMBER OF MEN WHO DIED OF LUNG CANCER, AGE-STANDARDIZED
DEATH RATES PER 100,000 MAN-YEARS AND MORTALITY RATIOS,
BY TYPE OF SMOKING (LIFETIME HISTORY) AND AGE AT START
OF STUDY.

Type of smoking (lifetime history)	Age 35-54		Age 55-69		Age 70-84		All ages 35-84	
	No. of Deaths	Death Rate	No. of Deaths	Death Rate	No. of Deaths	Death Rate	No. of Deaths	Death Rate
Never smoked regularly	21	7	41	19	21	35	83	13
Pipe only	2	5	17	50	15	110	34	29
Pipe and cigar	1	2	9	24	11	81	21	16
Cigar only	6	14	27	46	9	54	42	28
Cigarette & other	145	45	355	167	98	295	598	107
Cigarette only	427	57	743	216	115	311	1285	131
Total	602	41	1192	132	269	155	2063	81

Lung Cancer Mortality Ratios (Men)

Type of smoking	Age 35-54	Age 55-69	Age 70-84	All ages 35-84
Never smoked regularly	1.00	1.00	1.00	1.00
Pipe only	0.71	2.63	3.14	2.23
Pipe and cigar	0.29	1.26	2.31	1.23
Cigar only	2.00	2.42	1.54	2.15
Cigarette & other	6.43	8.79	8.43	8.23
Cigarette only	8.14	11.37	8.89	10.08

Lung cancer death rates increase with degree of exposure as measured by each of the three indices. For example, the lung cancer mortality ratio (ages 35-84) increased from 4.62 for men who smoked one to nine cigarettes a day up to 18.77 for men who smoked 40 or more cigarettes a day.

In a number of different studies, it has been found that lung cancer death rates are lower among former cigarette smokers who gave up the habit, than among men who continue to smoke cigarettes. One would like to know how soon the risk begins to diminish after the cessation of smoking. This is difficult to ascertain for three reasons:

1) In some instances, symptoms produced by undiagnosed lung cancer may lead a man to stop smoking; and it seems unlikely that giving up smoking would lead to the regression of an already established carcinoma.

2) Unless the number of ex-smokers under observation is extremely large, subjects must be traced for several years to accumulate enough person-years of exposure to risk for lung cancer death rates to be reasonably stable statistically.

3) Many smokers give up the habit for a few months or a year or two and then resume smoking again (16).

Table 3 is confined to men aged 50 to 74 (at the time of enrollment) who either had a lifetime history of only cigarette smoking or who had never smoked regularly. At the time of enrollment, a few subjects said that they had lung cancer; these are excluded. The men are divided into three groups according to their status at the time of enrollment: those who were currently smoking, those who had stopped smoking, and those who had never smoked regularly. The ex-smokers are divided by years since last smoking and by former amount of smoking. The current smokers are divided by current amount of smoking.

TABLE 2 -- NUMBER OF LUNG CANCER DEATHS, AGE-STANDARDIZED DEATH RATES AND MORTALITY RATIOS BY CURRENT NUMBER OF CIGARETTES SMOKED PER DAY, DEGREE OF INHALATION AND AGE BEGAN SMOKING. (FIGURES FOR MEN WHO NEVER SMOKED REGULARLY ARE SHOWN FOR COMPARISON).

Number of cigarettes a day, degree of inhalation, and age began smoking	Age 35-54		Age 55-69		Age 70-84		All ages 35-84	
Current No. of cigarettes a day	No. of Deaths	Death Rate	No. of Deaths	Death Rate	No. of Deaths	Death Rate	No. of Deaths	Death Rate
1-9	14	37	24	90	5	94	43	60
10-19	35	37	95	189	21	333	151	112
20-30	267	74	390	318	44	502	701	191
40 +	67	80	94	399	9	788	170	244
Degree of inhalation								
None	9	34	48	212	7	150	64	104
Slight	29	41	86	210	15	260	130	116
Moderate	224	67	338	286	33	418	595	170
Deep	120	79	131	319	23	848	274	221
Age began smoking								
25 +	11	24	28	99	3	60	42	53
20-24	68	50	117	228	11	303	196	131
15-19	217	68	326	310	45	581	588	191
<15	72	105	97	345	16	491	185	218
Never smoked regularly		7		19		35		13

<u>Lung Cancer Mortality Ratios (Men)</u>

Current No. of cigarettes a day				
1-9	5.29	4.74	2.69	4.62
10-19	5.29	9.95	9.51	8.62
20-30	10.57	16.74	14.34	14.69
40 +	11.43	21.00	22.51	18.77
Degree of inhalation				
None	4.86	11.16	4.19	8.00
Slight	5.86	11.05	7.43	8.92
Moderate	9.57	15.05	11.94	13.08
Deep	11.29	16.79	24.23	17.00
Age began smoking				
25 +	3.43	5.21	1.71	4.08
20-24	7.14	12.00	8.66	10.08
15-19	9.71	16.32	16.60	14.69
<15	15.00	18.16	14.03	16.77
Never smoked regularly	1.00	1.00	1.00	1.00

From data obtained in repeat questionnaires we
know that some of the men here classified as
ex-smokers resumed smoking at a later date (this
being most frequent among men who had stopped less
than two years before enrollment in the study) (16).
We also know that some of those here classified as
current smokers gave up the habit at a later date.
These changes in habits have not been taken into
consideration because we lack information on such
changes as may have occurred among men who died
during the two-year intervals between repeat
questionnaires.

The lung cancer death rates for men who had
given up cigarette smoking less than one year before
enrolling in the study were about the same as for men
who were currently smoking cigarettes at that time.
Those who had given up the habit for more than one
year had lower lung cancer death rates than the
current smokers. Among ex-cigarette smokers, lung
cancer death rates decreased with length of time
since last smoking.

Findings in Females

In the United States, cigarette smoking became a
popular habit among men some years before it started
to become popular among women. During the period
covered by our study, few women in the older age
groups were cigarette smokers; and fewer young women
than young men were cigarette smokers. As a group,
the female smokers had taken up the habit later in
life than the male smokers, smoked fewer cigarettes a
day, tended to inhale the smoke less deeply and were
more likely to smoke low-tar, low-nicotine cigarettes
(17).

Table 4 shows lung cancer death rates of female
subjects classified by their smoking habits. Figures
shown on the line labeled "history of smoking"
include ex-smokers as well as current smokers. On
lower lines, current smokers are classified by three
different indices of exposure: number of cigarettes
smoked per day, degree of inhalation of cigarette
smoke, and age they began cigarette smoking. Lung
cancer death rates are higher in the smokers than in
the non-smokers and increase with amount of cigarette
smoking.

TABLE 3 -- AGE STANDARDIZED LUNG CANCER DEATH RATES FOR EX-CIGARETTE SMOKERS WITH A HISTORY OF CIGARETTE SMOKING ONLY, BY FORMER NUMBER OF CIGARETTES SMOKED PER DAY, AND YEARS SINCE LAST CIGARETTE SMOKING. DEATH RATES FOR CURRENT CIGARETTE SMOKERS WITH A HISTORY OF CIGARETTE SMOKING ONLY AND MEN WHO NEVER SMOKED REGULARLY ARE SHOWN FOR COMPARISON. MEN AGED 50-74 WHO DID NOT HAVE A HISTORY OF LUNG CANCER AT THE START OF THE STUDY.

Ex-cigarette smokers (years since last cigarette smoked)	Smoked 1-19 cigarettes a day			Smoked 20 + cigarettes a day		
	No. of Men	No. of Deaths	Death Rate	No. of Men	No. of Deaths	Death Rate
Under 1 year	812	5	114	2,308	32	283
1-4 years	1,990	6	53	5,662	43	162
5-9 years	1,913	2	20	6,108	30	104
10 + years	4,638	2	7	8,681	13	29
Total ex-smokers	9,353	15	28	22,759	118	101
Current cigarette smokers	24,523	154	120	58,739	690	271
Never smoked regularly	62,590	60	16	62,590	60	16

The lung cancer death rates and mortality ratios shown in Table 4 for women are not as high as those shown in Table 2 for men. This is almost certainly due in part to difference in exposure. For example, among men and women who smoked the same number of cigarettes per day, the women, as a group, inhaled the smoke less deeply than the men and had started smoking later in life. This difference in habits between the sexes is more pronounced in older than in younger age groups. However, it probably does not fully account for the sex difference in lung cancer death rates.

Other Evidence

Evidence from histologic studies carried out on men who died and came to autopsy is fully consistent with the evidence from epidemiological studies (18,19). Cigarette smoking is associated with the following changes in bronchial epithelium: loss of cilia in many areas, hyperplasia, squamous metaplasia, a great increase in the number of cells with atypical nuclei, and the occurrence of carcinoma-in-situ. (See Chapter 3.) All of these changes occur more frequently in cigarette smokers than in non-smokers, and increase in frequency with amount of cigarette smoking. Such changes are found to a far lesser extent in former cigarette smokers who gave up the habit some years prior to their terminal illness, than in men who continued to smoke cigarettes up to the time of their terminal illness (20).

Experimental studies have shown that exposure to cigarette smoke inhibits the action of cilia of the bronchial epithelium (21). This reduces the efficiency of removal of foreign material from the bronchial tubes.

Many investigators have produced skin cancer in experimental animals by the application of cigarette smoke condensates (22); and invasive lung tumors (including early squamous cell carcinoma) have been produced in beagle dogs by the smoking of non-filter cigarettes daily for over two years (23,24). Such experiments have been carried out primarily as a means of testing the relative carcinogenicity of various types of cigarettes and various components of

cigarette smoke. Hopefully, it may be possible to
develop cigarettes which are less potent in respect
to the production of lung cancer than are cigarettes
which are most popular at the present time.

URBAN AIR POLLUTION

It is well established that occupational
exposure to certain specific types of air
contaminants carries a greatly increased risk of lung
cancer. Here we are primarily concerned with urban
air pollution to which all residents of modern cities
and metropolitan areas are exposed to a greater or
lesser degree. However, we cannot altogether avoid
the subject of occupational exposure since, in
developed countries, a considerable proportion of all
men (including many farmers) are occupationally
exposed to air contaminants of one sort or another.

Cancer produced by exposure to a chemical agent
typically does not occur until long after initial
exposure; and, unless the agent is retained in the
body, the risk of developing the disease generally
declines if exposure is discontinued. Therefore, the
most valid design for investigating the association
between exposure to a specified substance and the
occurrence of cancer requires knowledge of the
lifetime history of the exposure of each subject --
at least a rough estimate of the time since first
exposure and a rough estimate of degree of exposure.
As previously described, such information was
obtained in both retrospective and prospective
studies of exposure to tobacco smoke. It has also
been obtained with a fair degree of accuracy in many
studies of occupational exposure to specific agents.
Unfortunately, it would be extremely difficult, if
not impossible, to obtain such information on each
subject included in a study of general urban air
pollution. There are manifold problems:

> 1) "Urban air pollution" is a non-specific
> term in the sense that it covers an
> extremely wide range of different types
> of pollution: various gases, organic
> particles, inorganic particles, and even
> particles too large to be inhaled.
> There is probably no urban area in which
> just one type of air pollutant is

TABLE 4 -- LUNG CANCER (WOMEN). NUMBER OF DEATHS, AGE STANDARDIZED DEATH RATES AND MORTALITY RATIOS, BY TYPE OF SMOKERS (LIFETIME HISTORY), CURRENT NUMBER OF CIGARETTES SMOKED PER DAY, DEGREE OF INHALATION AND AGE BEGAN SMOKING; BY AGE AT START OF STUDY.

Smoking history	Age 40-54		Age 55-74		All Ages 40-74	
	No. of Deaths	Death Rate	No. of Deaths	Death Rate	No. of Deaths	Death Rate
Never smoked regularly	45	4	121	13	166	8
History of cigarette smoking	94	14	67	31	161	21
Current regular cigarette smoking						
Current No. of cigarettes a day						
1-9	7	6	8	14	15	10
10-19	18	10	16	31	34	19
20-39	51	21	34	63	85	39
40 +	10	53	2	69	12	60
Degree of inhalation						
None	7	10	11	23	18	16
Slight	17	13	13	25	30	18
Moderate	40	13	26	49	66	28
Deep	21	26	9	99	30	57
Age began smoking						
25 +	11	6	40	34	51	18
20-24	24	16	10	42	34	27
15-19	43	20	9	67	52	40
< 15	6	36	-	-	6	20

	Lung Cancer Mortality Ratios (Women)		
Never smoked regularly	1.00	1.00	1.00
History of cigarette smoking	3.50	2.38	2.63
Current regular cigarette smoking			
Current No. of cigarettes a day			
1-9	1.50	1.08	1.25
10-19	2.50	2.38	2.38
20-39	5.25	4.85	4.88
40 +	13.25	5.31	7.50
Degree of inhalation			
None	2.50	1.77	2.00
Slight	3.25	1.92	2.25
Moderate	3.25	3.77	3.50
Deep	6.50	7.62	7.13
Age began smoking			
25 +	1.50	2.62	2.25
20-24	4.00	3.23	3.38
15-19	5.00	5.15	5.00
< 15	9.00	-	2.50

present, and probably no two areas with precisely the same qualitative and quantitative combination of pollutants.

2) Within the same metropolitan area, the type and amount of pollution varies in different neighborhoods, and varies from day to day and year to year. Furthermore, qualitative and quantitative analyses of air samples have not been carried out on a routine basis in many localities until recent years. Even now, one may question the adequacy of such sampling for determining the current exposure of individuals living in different neighborhoods of the same metropolitan area.

3) A large proportion of American men live at some distance from their place of work. They may be exposed to different types and degrees of air pollution at home, on their way to work, and in their place of work.

4) Americans are remarkably mobile; many move from one location to another every few years. This complicates the problem of ascertaining the type and extent of exposure. Furthermore, state-of-health can influence whether a person moves from one location to another. This is an additional complicating factor.

Because of these difficulties, we are generally unable to obtain an accurate estimate of the degree of exposure of an individual to each of various types of air pollutants during his lifetime. As a poor substitute, we can divide individuals into groups by residence history and use this as a very crude index of exposure history. Alternatively, we can ascertain lung cancer death rates in various localities which currently differ in type or degree of air pollution. In some instances, a compromise may be made between these two procedures. Whichever procedure is used, smoking habits and occupational exposure should be taken into consideration.

Table 5, based upon data from the American Cancer Society's prospective study previously described, is confined to male subjects who, at the time of enrollment, said that they had lived in their present neighborhood for at least ten years. Thus they had a minimum of ten years of exposure to the type and amount of air pollution occurring in their neighborhoods during those years. The total group was divided into six subgroups: men who never smoked regularly, and five sets of smokers classified by type and amount of smoking. Lung cancer death rates were calculated for men in each five-year group, in each of the six subgroups. These death rates were applied to the man-years of exposure to risk of men in each of the various categories shown in Table 5. The resulting figures represent the expected number of lung cancer deaths in each category, adjusted for age distribution and smoking habits. The observed number of lung cancer deaths divided by the expected number yields the mortality ratio. By definition, the mortality ratio for all subjects combined is 1.00.

The subjects are divided into various groups by place of residence. Within each of these groups they are subdivided according to whether they said that they were or ever had been occupationally exposed to dust, fumes, vapors, gases, or X-rays. The exposures reported covered a wide range (e.g., firemen exposed to smoke, garage workers exposed to automobile exhausts, asbestos workers, miners, farmers exposed to insecticide sprays, etc.) Many of the exposed men probably had only a low level of occupational exposure for a relatively short length of time. Others may have had a high level of exposure for many years.

Without regard to place of residence, the lung cancer mortality ratio was 1.09 for men with occupational exposure and 0.96 for men without occupational exposure, a relative difference of 13.5%. In large metropolitan areas, the relative difference between the occupationally exposed and unexposed groups was 26%, in smaller metropolitan areas 18%, and in non-metropolitan areas 7%. These differences are probably due to different types of occupational exposures in different areas.

TABLE 5 -- OBSERVED AND EXPECTED NUMBER OF LUNG CANCER DEATHS BY PLACE OF RESIDENCE AND BY OCCUPATIONAL EXPOSURE TO DUST, FUMES, GASES, OR X-RAYS. ADJUSTED FOR AGE AND FOR SMOKING HABITS. CONFINED TO MEN WHO HAD LIVED IN SAME NEIGHBORHOOD FOR LAST 10+ YEARS.

Place of residence	Occupationally exposed to dust, fumes, etc.			Not occupationally exposed to dust, fumes, etc.		
	Obs. No.	Exp. No.	Ratio	Obs. No.	Exp. No.	Ratio
Total, all male subjects	576	530.5	1.09	934	979.7	0.96
Metropolitan area, pop. 1,000,000 +	165	134.1	1.23	281	285.7	0.98
City	92	69.1	1.33	168	158.3	1.06
Town or Rural	73	65.0	1.12	113	127.4	0.89
Metropolitan area, pop. <1,000,000	166	145.4	1.14	271	280.5	0.97
City	92	83.3	1.10	170	184.0	0.92
Town or Rural	74	62.1	1.19	101	96.5	1.05
Non-metropolitan area:	245	251.0	0.98	382	413.5	0.92
Town	102	104.9	0.97	200	199.1	1.00
Rural	143	146.1	0.98	182	214.4	0.85
Los Angeles, Riverside, & Orange Counties, Cal.	30	21.9	1.37	38	39.6	0.96
Farmers	63	77.6	0.81	71	92.9	0.76
8 Cities: High particulates (130-180 $\mu g/m^3$)	45	32.9	1.37	66	73.9	0.89
11 " Moderate " (100-129 $\mu g/m^3$)	21	18.8	1.12	39	49.5	0.79
14 " Low " (35-99 $\mu g/m^3$)	48	37.4	1.28	110	100.1	1.10
9 Cities: High Benz. Sol. (8.5-15.0 $\mu g/m^3$)	28	21.0	1.33	52	51.5	1.01
10 " Moderate " (6.5- 7.9 $\mu g/m^3$)	44	32.7	1.35	65	75.1	0.87
12 " Low " (3.4- 6.3 $\mu g/m^3$)	33	29.2	1.13	76	81.8	0.93

Clearly, occupational exposure should be taken into consideration in any study of the possible effects of general urban air pollution. The simplest way of doing this is to confine attention to men without occupational exposure.

Men Without Occupational Exposure

In the top part of Table 5, the subjects are divided into six groups by size of place of residence according to the 1960 census of the United States. The term "metropolitan area" means a county or a group of contiguous counties with at least one city of 50,000+ inhabitants, or "twin cities" with a combined population of at least 50,000. As used here, the term "town" means a place with a population of 2,500 to 49,999 people, and "rural" means those who live in the country or a village with less than 2,500 people. In some metropolitan areas, legally independent towns abut on a central city and, in a non-legal sense, are actually a part of the city (a situation similar to Greater London in contrast to the City of London).

Generally speaking (but with exceptions), it may be assumed that urban air pollution tends to be greater in large metropolitan areas than in smaller metropolitan areas, far less in non-metropolitan areas, and least in rural parts of non-metropolitan areas. Most of the non-metropolitan areas included in this study are far removed from any city and many do not even contain a large town.

Among men without occupational exposure, the lung cancer mortality ratio was 0.98 for those living in large metropolitan areas (1,000,000+ population), 0.97 for those living in smaller metropolitan areas, and 0.92 for those living in non-metropolitan areas. The highest mortality ratios (1.06 and 1.05) were for men living in cities in large metropolitan areas, and for men living in towns and rural parts of smaller metropolitan areas. The lowest mortality ratios (0.85 and 0.89) were for men living in rural parts of non-metropolitan areas, and for men living in towns and rural parts of large metropolitan counties. The mortality ratio for men living in towns in non-metropolitan areas (1.00) was higher than the mortality ratio of men living in cities in smaller metropolitan areas (0.92). This set of figures gives little or no support to the hypothesis that urban air pollution has an important effect upon lung cancer death rates.

Los Angeles county in California, and major parts of two adjacent counties (Riverside and Orange), have unusually heavy air pollution in respect to oxidants and carbon monoxide (13). They also have high air pollution in terms of total suspended particulate matter and benzene-soluble particulate matter. The lung cancer mortality ratio for men living in these three counties was the same as for all subjects without occupational exposure (0.96).

Data are shown for farmers, including retired farmers living in towns, but excluding: 1) farmers living in metropolitan areas of 500,000+ population and 2) retired farmers living in cities or in metropolitan areas of 500,000+ population. The majority of these farmers lived in strictly rural areas far from any large city -- and far from any major medical center. Their lung cancer mortality ratio was only 0.76. We suspect that this figure is artificially low for two reasons:

1) In strictly rural areas of the United States there are usually few doctors and usually little in the way of medical facilities. Under such conditions, some deaths due to lung cancer may be mistakenly attributed to other causes.

2) In past times, when a farmer's health began to fail, he usually remained on the farm and his son took over the work. Today, he is far more likely to move to a city. This selective removal from rural areas of men in ill health reduces the death rate in rural areas.

Data on the mean level of suspended particulate matter in the air of 57 American cities is provided in Statistical Abstracts in the United States, 1970 (25), for the year 1968. The mean levels ranged from 32 mg/m^3 to 306 mg/m^3. It is likely that the mean level in some of the cities changed considerably during the last four decades or so. However, lacking evidence to the contrary, we will assume that the rank order of these cities in respect to suspended particulate matter did not change greatly. Thirty-three of the cities were included in our study and we divided them into three groups by mean level of suspended particulate matter in 1968.

The mortality ratios (for men without occupational exposure) were: 0.89 for cities with the highest mean levels of suspended particulate matter; 0.79 for cities within the intermediate category; and 1.10 for cities with the lowest mean levels of suspended particulate matter. Since it seems unlikely that suspended particulate matter decreases the risk of lung cancer, we conclude that urban air pollution as measured by this index is unrelated to death rates from lung cancer.

Statistical Abstracts of the United States, 1970 (25) also provides information on the mean level of benzene-soluble organic matter for the year 1968 in the air of 55 cities, 31 of which were included in the study. As shown in Table 5, there appears to be little if any association between lung cancer mortality ratios and this index of urban pollution.

CONCLUSION

In a review of the literature published some years ago (26), the authors concluded that there was no firm evidence in support of the hypothesis that general urban air pollution increases the risk of lung cancer to an important degree, if at all. Data from our study supports that conclusion; and we are unaware of any evidence which convincingly leads to a contrary conclusion.

Available evidence does not rule out the possibility that general urban air pollution may perhaps lead to a slight increase in the risk of lung cancer. It also does not rule out the possibility that if no efforts were made to control air pollution, then at some future date it might increase to a level such that it would result in a significant increase in the risk of lung cancer. Fortunately, for good and sufficient reasons (other than lung cancer risk), steps are now being taken to reduce air pollution. If reasonably successful, these steps should eliminate the possibility that general urban air pollution will result in an increase in risk of lung cancer at some future date.

A WORD OF CAUTION IS IN ORDER -- Up until now, this discussion has been confined to the subject of general urban air pollution to which all persons living in metropolitan areas are more or less exposed. But people who live in the neighborhood of an industrial plant which discharges a specific type of air pollutant in considerable quantities may be in a different position. Some such pollutants (e.g. asbestos dust) greatly increase the risk of lung cancer among occupationally exposed workers; and it is possible that, in some instances, exposure of people living in the vicinity of a plant may reach dangerous levels. This matter deserves more attention than it has received in the past.

REFERENCES

1. Public Health Service, U. S. Department of Health, Education, and Welfare. (1971). The Health Consequences of Smoking. Supplement to the 1967 Public Health Service Review. Washington, D. C.

2. Royal College of Physicians of London. (1971). Smoking and Health Now. Pitman Publishing Corporation, London.

3. Müller, F. H. (1939). Tabakmissbrauch und Lungencarcinom. Z. Krebsforsch 49: 59.

4. Ochsner, A. and De Bakey, M. (1939). Primary pulmonary malignancy treatment of total pneumonectomy: Analysis of 79 collected cases and presentation of 7 personal cases. Surg. Gynec. & Obst. 68: 435.

5. Wynder, E. L. and Graham, E. A. (1950). Tobacco smoking as a possible etiologic factor in bronchiogenic carcinoma: A study of 684 proved cases. J. Amer. Med. Assoc. 143: 329.

6. Levin, M. L., Goldstein, H. and Gerhardt, P. R. (1950). Cancer and tobacco smoking: A preliminary report. J. Amer. Med. Assoc. 143: 336.

7. Doll, R. and Hill, A. B. (1952). A study of the aetiology of carcinoma of the lung. Brit. Med. J. 2: 1271.

8. Doll, R. and Hill, A. B. (1954). The mortality of doctors in relation to their smoking habits: A preliminary report. Brit. Med. J. 1: 1451.

9. Hammond. E. C. and Horn, D. (1954). The relationship between human smoking habits and death rates: A follow-up study of 187,766 men. J. Amer. Med. Assoc. 155: 1316.

10. Best, E. W. R., Josie, G. H. and Walker, C. B. (1961). A Canadian study of mortality in relation to smoking habits: A preliminary report. Canad. J. Pub. Health 52: 99.

11. Dorn, H. F. (1958). The mortality of smokers and non-smokers. Proc. Soc. Stat. Sect., Amer. Stat. Assoc. , 34.

12. Dunn, J. E., Linden, G. and Breslow, L. (1960). Lung cancer mortality experience of men in certain occupations in California. Amer. J. Pub. Health 50: 1475.

13. Buell, P., Dunn, J. E. and Breslow, L. (1967). Cancer of the lung and Los Angeles-type air pollution. Prospective Study. Cancer 20: 2139.

14. Hammond, E. C. (1964). Smoking in relation to mortality and morbidity. Findings in the first thirty-four months of follow-up in a prospective study started in 1959. J. Nat. Cancer Inst. 32: 1161.

15. Hammond, E. C. (1966). Smoking in relation to the death rates of one million men and women. Monograph 19, National Cancer Institute, Dept. Health, Education, and Welfare, Washington, D. C., 129.

16. Hammond, E. C. and Garfinkel, L. (1964). Changes in cigarette smoking. J. Nat. Cancer Inst. 33: 49.

17. Hammond, E. C. and Garfinkel, L. (1961). Smoking habits of men and women. J. Nat. Cancer Inst. 27: 419.

18. Auerbach, O. et al. (1961). Changes in bronchial epithelium in relation to cigarette smoking and in relation to lung cancer. New Eng. J. Med. 265: 253.

19. Auerbach, O. et al. (1962). Changes in bronchial epithelium in relation to sex, age, residence, smoking, and pneumonia. New Eng. J. Med. 267: 111.

20. Auerbach, O. et al. (1962). Bronchial epithelium in ex-cigarette smokers compared with current cigarette smokers and non-smokers. New Eng. J. Med. 267: 119.

21. Hilding, A. C. (1956). On cigarette smoking, bronchial carcinoma and ciliary action. New Eng. J. Med. 254: 775.

22. Wynder, E. L. and Hoffman, D. (1967). Tobaco and Tobacco Smoke: Studies in Experimental Carcinogenesis. Academic Press, New York.

23. Hammond, E. C. et al. (1970). Effects of cigarette smoking on dogs. I. Design of experiment, mortality, and findings in lung parenchyma. Arch. Env. Health 21: 740.

24. Auerbach, O. et al. (1970). Effects of cigarette smoking on dogs II. Pulmonary neoplasms. Arch. Env. Health 21: 754.

25. U. S. Bureau of the Census. (1970). Statistical Abstract of the United States: 1970. (91st annual ed.). Washington, D. C.

26. Cornfield, J. et al. (1959). Smoking and lung cancer: Recent evidence and discussion of some questions. J. Nat. Cancer Inst. 22: 173.

CHAPTER 13. OCCUPATIONAL
LUNG DISEASES

IRVING J. SELIKOFF, Mount Sinai School of Medicine,
City University of New York, New York

Lung disease as a consequence of inhalation of dusts, fumes, and vapors in occupational circumstances has a fairly long and well-documented history, with references to other materials in 1713 by Ramazzini (1), with sturdy references to mining dusts in 1556 by Agricola (2), and even earlier references if we rummage through ancient writings. In the succeeding centuries these initial observations were confirmed and extended (3,4), so that thirty years ago chest physicians considered that they had a well-defined, albeit limited, compartment for "occupational chest diseases", meant primarily for those conditions resulting from the inhalation of fibrogenic dusts, and particularly those dusts containing free silica.

This tidiness has been upset. Matters have become much more complicated. Not only has the spectrum of etiologic agents been greatly widened, but their effects are observed to be much more varied, their mechanisms far more sophisticated, and their importance considerably greater. Not only are fairly limited employed groups, often sequestered in geographically isolated areas, potentially at risk, but large numbers of working people and, in some instances, even the community at large. This last possibility -- environmental lung disease associated with agents usually met with in occupational circumstances -- has particular interest, and will be later considered.

OCCUPATIONAL LUNG DISEASES — 1971

Historical Status of Silica

The elucidation of the special pathogenetic potential of crystalline silica, and its recognition from among a large variety of ill-defined mineral dusts, was a major scientific achievement (5). In a sense, though, we then became victims of our own success; knowledge of silica's potential contrasted with lack of information concerning the biological effects of other dusts. Such discrepancy came to be equated with the concept that, by and large, the hazard of an inhaled dust could largely be related to its crystalline silica content.

This has now changed. The silica pneumoconioses remain (silicosis), but much that is new has been added. Perhaps a recounting of the range of such additions would be more instructive than a simple catalogue of a list of substances now known to be potentially associated with disease, especially since this list is still expanding.

Parenthetically, it may be noted that, in comparison with a number of other substances now being studied, silica does not turn out to be extraordinarily fibrogenic after all, and if only small amounts are present in human lung, in either occupational or environmental circumstances, overt disease is usually absent (6).

Silicates

A number of fibrous and non-fibrous silicates have been identified as causing occupational lung disease. The most important are the fibrous silicates generically known as asbestos. These include several fibrous minerals (chrysotile, amosite, crocidolite, anthophyllite, tremolite, and others) which differ chemically and structurally. With their wide use in industry, they now constitute a major problem. Information concerning the number of workmen occupationally exposed to these fibrous minerals in the United States is hard to come by. Perhaps 200,000 would be a close approximation. Nor are data yet available which would tell us the risk to which these people are exposed in each of various trades involved. For one group, however, the risk has been measured. Among asbestos insulation workers in the United States, serious lung disease has been demonstrated; approximately 1 in 5 deaths among these men, in one geographical area, has been found to be due to lung cancer, 1 in 10 to asbestosis, 1 in 10 to gastrointestinal cancer, and 1 in 10 to pleural and peritoneal mesothelioma (7,8). Prevalence studies in the same group show that very few escape scarring of the lung and/or pleura. Nine of ten men show such changes on chest roentgenogram when examined thirty or forty years from onset of employment (9).

Talcosis, the pneumoconiosis associated with the inhalation of talc, may, in part, be considered a variant of asbestosis, since many of the talc workers so afflicted have inhaled material containing such fibrous silicates as chrysotile and tremolite. This would include both miners and millers of the material, as well as those who use it in industry as, for example, some rubber workers. On the other hand, other talcs have little or no fibrous component, being primarily composed of a platy mineral; it is not yet known whether the clinical syndromes of talcosis (10,11) are equally associated with the platy varieties. They may well be, since mica, also a platy silicate, is associated with pneumoconiosis. The matter is not yet clear since a number of the mica ores contain free silica in significant proportion (12).

Non-fibrous silicates may also cause pulmonary disease. This has long been known, for example, with Fuller's Earth (largely montmorillonite) and with kaolin (13). Even so apparently inert a mineral as nephaline has been reported to cause fatal lung disease, in an instance with massive exposure (14).

The wide variety of silicates now known to be associated with human disease is of considerable interest, since so much of the earth's crust is silicate in nature, and dissemination of silicate particles into the environment should be a matter for careful study.

Coal Workers' Pneumoconiosis

Gradually, order and understanding are evolving in the complex problem of lung disease among coal workers, both miners and other coal handlers (15,16). Difficulties have included the variable nature of coal itself, the frequent disparity between x-ray findings and clinical disability, the simultaneous operation of other causes of emphysema, and the effect of concomitant silica, cigarette smoking, and non-specific respiratory infection. To some extent, the weight to be given to each of these and other problems, is still a matter of study and debate. But however the final adjudication is made, the sum total adds up to a major lapse in occupational medicine.

There are few accurate data on the total number of underground coal miners who worked in this country during this century; employment, recently as low as 125,000 men, at times reached over 500,000. A large proportion of regularly employed coal miners may expect, if they spend their lifetime in this occupation, to have evidence of lung abnormality. Sometimes this abnormality is primarily radiological, with little disability or risk of premature death. Sometimes, especially when progressive massive fibrosis supervenes (17), death from respiratory insufficiency and cor pulmonale occurs. Since these events occur two, three, or more decades after the onset of exposure, and since our inattention to this problem in the United States until the 1960's provides little baseline information, the incidence of serious disability or work-related death is not yet precisely known. It is expected that studies in progress will provide useful information in the next several years, which might then be compared with the much more adequate information available in Great Britain (18).

Mixed-dust Pneumoconiosis

One starts by studying inhaled materials as though they occur in unique isolation, since it would otherwise be difficult to decipher the pathogenic mechanisms and potential hazard of the materials in question. In occupational circumstances, however, such unique exposures are uncommon. For that matter, even the single minerals are usually composites. For example, chrysotile as received from the mine often contains such amphiboles as anthophyllite or tremolite. The mixed nature of talc has already been mentioned. Coal is an excellent example of such a mixture. Freshly mined diatomaceous earth, depending on the seam from which it is derived, may contain measurable quantities of crystalline silica, in addition to the amorphous form.

Commercial use complicates matters further. Thus, for example, an insulation worker has the opportunity of inhaling not only asbestos (generally chrysotile in the United States, but sometimes amosite), but also fibrous glass, diatomaceous earth, calcium silicate, cement, etc., together with such trace elements as may have been added during manufacture of the insulating material.

There has been increasing realization that the reason why "classical" clinical and x-ray findings of specific pneumoconioses are not always seen, is that complex rather than "pure" exposures are common or, in some circumstances, probably the rule. An important condition in which this is true is foundry workers' lung disease, where crystalline silica, silicates, metal dusts, and high temperatures may all combine to modify the effects to be expected from any one of these agents. When we add to these the additional powerful influence of cigarette smoking, the blurring of diagnostic outlines can be appreciated (19). Very little is known of the number of individuals who might be affected by mixed exposures, and still less is known of the pathogenetic implications of the interactions among two or more dusts simultaneously present in human lung. They, too, form an area of inquiry of considerable interest, since environmental lung disease is likely to be far more frequently associated with the presence in lung of a variety of substances rather than one or two.

Metals

The effect of metals in the lung is also being critically reexamined. Once thought to be largely limited to radiologically striking but clinically benign effects of tin (stannosis) (20) and iron (siderosis), or to the generally self-limited metal fume fever from zinc, copper, antimony, etc. (21), the horizons of our concern are now much wider. In part, this has resulted from the clear definition of the importance of beryllium lung disease (22), sometimes resulting in severe or fatal lung scarring, and wartime experiences with aluminosis (23).

In recent years, clinical lung disease has been demonstrated from chromates (lung cancer) (24), nickel (bronchopneumonia with nickel carbonyl as well as lung cancer) (25), bauxite (Shaver's disease) (26), and tungsten and cobalt "hard metal disease" (27). Interest in these clinical entities is heightened by current concern with trace elements in community air, and it is likely that study of occupational groups exposed to these specific elements will provide useful information.

"Inert" Dusts

We have long had a category of "nuisance" or "inert" dusts, which were presumed to be essentially harmless. The list is becoming smaller, and those remaining should be given only provisional clearance. One problem has been that the toxicological potential of such "inert" dusts has had to be measured largely by the results of animal studies. These only imperfectly resemble the human situation. Inhalation experiments with iron dust fail to demonstrate the lung cancer of hematite miners, and the paucity of successful lung cancer induction experiments in animals contrasts sharply with the frequency of lung cancer in asbestos workers who smoke cigarettes (8). Below, attention is called to such variables as multiple factor effects, the necessity for a long-lapse period, and species and tissue specificity, all difficult to encompass in animal studies. Similarly, particles might themselves be "inert" but might act as carriers for other particles (28), gases (29), radioactive substances. The inclusion of a wide variety of materials under a categorical name may obscure subtle but important differences. Fibrous glass is a case in point. There is no such single entity, but rather a wide range of fibrous glass materials, differing structurally, chemically, physically, and dimensionally, with various binders, additives, or coatings incorporated. Observations made with one sample are not necessarily applicable to another.

Even pure carbon dust might, under some circumstances, be a hazard (30), if inhaled in large enough quantities, over a long enough period of time; this, without taking into account the possibility that carbon black currently is usually derived from the burning of natural gas, with incompletely studied residuals of polycyclic hydrocarbons. When we step a bit away from the "pure" material to, for example, graphite, lung disease is found (31).

Chemical Agents

There is considerable knowledge concerning direct toxic effects of a large number of chemicals on the tracheobronchial tree and the pulmonary parenchyma. Catalogued in toxicological texts, these effects have been noted primarily on the basis of their acute irritative reactions and on appropriate animal exposure studies. There is little information concerning long-term effects of repeated, or low-level exposures, nor is much known of the pulmonary disease potential of a large variety of newly introduced chemicals, made available by industry during the past two decades. That the effect of some such exposures may be unique can be inferred from our experiences with isocyanates (32) and, perhaps, proteolytic enzymes (33). This subject has hardly been given the critical attention it deserves, in view of the large number of workmen involved throughout American industry. A recent unhappy example has been the demonstration of excess lung cancer among workmen exposed to coke oven fumes in the steel industry (34).

Organic Dusts

Occupational lung disease resulting from the inhalation of organic dusts is now appreciated to be an important clinical entity. As yet, there has been inadequate analysis of the mechanisms involved; perhaps it will be found that these mechanisms are as varied as the etiological agents involved.

Important examples include respiratory disease among some cotton textile workers (byssinosis) (35), farmer's lung (36), lung disease among detergent enzyme workers (38), bagassosis (38), and the variety of conditions grouped under the rubric of extrinsic allergic alveolitis (39) (as farmer's lung, mushroom workers' lung, suberosis, lung disease among paprika workers, sisal workers, coffee workers, and workmen in our redwood forests on the West Coast). These conditions have exotic names; unfortunately, also, they may result in severe and even fatal pulmonary fibrosis.

For completeness, infectious lung disease, such as anthrax or coccidiodomycosis, contracted in occupational circumstances could be included, as could occupational asthmas.

Physical Agents

Physical insults add a fairly small number of additional cases of occupational lung disease. These include "nonspecific" respiratory disease seen among individuals exposed for long periods to high heat and low humidity, such as stationary firemen; those diseases seen with cold and high humidity, and the lung disease which seems to be increasing among firefighters in our cities. Smoke inhalation for the last may be a complex exposure; it can result in stubborn, disabling pulmonary disease. Caisson disease (40) from the liberation of nitrogen bubbles after exposure to high barometric pressure may be associated with lung cysts, and be complicated by the silicosis to which tunnel workers are subject. With the increasing use of hyperbaric medical facilities, it may be worthwhile to look for lung changes in medical personnel serving in these units.

An important occupational lung disease of physical origin is, of course, the lung cancer of uranium miners and other workmen exposed to the inhalation of radon daughters (41,42). Little is known of other effects on the pulmonary parenchyma of such radiation exposure, but the pulmonary fibrosis sometimes seen following radiation for breast cancer suggests that these should perhaps be looked at.

IMPORTANT VARIABLES

The pneumoconioses develop at the interface between living matter and the external world (43). As such, they provide a fruitful model for the study of many aspects of the larger problem of the biological effects of environmental alteration. Several advantages of such an occupational model are apparent: human populations are studied, often large enough to detect limited effects, long term prospective studies are feasible, and the environment can be measured and defined in relation to the effect considered. Perhaps more important, the complexities and variables so often inherent in human disease are included in the model.

Multiple Factors

In many occupational lung diseases, the disease process may be determined by influences other than the offending agent, which is a necessary but not a sufficient factor. The recognition of such interactions marks a useful advance in the last several years. Still incompletely studied, such interactions have already been identified in several important occupational lung diseases.

Pulmonary disability after silica exposure, for example, seems highly dependent upon whether or not the individual simultaneously smokes cigarettes (44), and the same will probably be found to be true for many instances of coal workers' pneumoconiosis as well. A clear example of such association is found in lung cancer from asbestos. Asbestos insulation workers who also smoke cigarettes have approximately 92 times the risk of dying of bronchogenic lung cancer, as men of the same age who neither smoke cigarettes nor work with asbestos (8); whereas their colleagues who do not smoke cigarettes have not been found to have greatly increased risk of bronchogenic lung cancer, despite considerable asbestos lung burden. Uranium miners, too, provide evidence for a similar multiple factor effect (45). Recently, lung disease among workmen in enzyme detergent factories has been found to occur more frequently among atopic individuals (37). Still unidentified constitutional factors may be responsible for the variation in individual response to fibrogenic dusts; after forty years of exposure, 5% of asbestos insulation workers still have normal chest x-rays (9).

Long Lapse Period

Some occupational lung diseases have an obvious cause-effect relationship; e.g.: acute chemical bronchitis, isocyanate reactions, metal fume fever, and some instances of beryllium lung disease. In other cases, the etiological association may not be so clear. This is true even of "acute silicosis," where several years may elapse between very heavy exposure and pulmonary disability. For many diseases, the lapsed period (I suggest that this term is preferable to "latent period," and would urge its adoption (46)) between onset of exposure and evidence of disease is very much longer; often two, three, four, or more decades. In our mobile society, where a single lifetime job is no longer necessarily the rule, prior exposure may be obscure and inadequately considered when a pulmonary complaint is evaluated. The matter becomes further complicated when we consider the influence of multiple factors; each may have its own lapsed period, starting at different times for the individual. The risk of lung cancer for an asbestos worker in his fiftieth year might be different if he began work at age 18 and cigarette smoking at 26 than if these two ages were reversed.

A World of Fine Particles

Some reorientation of our perspectives in considering dust diseases of the lung has occurred in the last several years, with the addition of electronmicroscopic studies in evaluation of airborne contaminants and lung dust. The prediction of Turkevich (47) seems to have been confirmed in many instances. It has been long known that submicron particles, in the Ångstrom unit range, are present in dust-contaminated air, but it was assumed that their dynamic behavior in the inhaled and exhaled airstreams would largely prevent their retention in the lung. For a number of dusts already studied, this is now found not to be the case (48), and it is likely that EM investigations will be profitable in many other occupational lung diseases. As a corollary, environmental studies will benefit from this extension of technique, especially when environmental contamination from occupational sources is under consideration.

Nucleation and Adsorption

Physical-chemical interaction among environmental agents is common, but has hardly begun to be studied in occupational lung disease. We do have scattered information on a number of relatively simple combinations, such as the adsorption of submicron particles onto the surface of other particles 1-5μ in diameter, for example. We have observed large numbers of chrysotile fibrils 350Å in diameter and 1000-2000Å in length, adsorbed on the surface of diatom particles, fibrous glass particles, or clay particles. Retention of chrysotile in the lung in such circumstances would depend much more on the behavior of their carrier particles than on their own dynamics. Radon daughters and mineral dust particles, nitrogen oxide, and carbon particles, provide examples of similar effects.

The problem is much more complex when one considers the particle-cell interface in tissue. Do the surfaces of glass particles in the lung affect kinin metabolism as in vitro observations suggest? Do such surface interactions bear on pathogenetic mechanisms?

These possibilities have therapeutic implications. One was thought of long ago; the alteration of the surface properties of silica by aluminum oxide in vitro led to the recommendation of aluminum dust treatment for silicotic miners. Recently, polyvinylpyridine-n-oxide has been found to block the effect of crystalline silica in the experimental animal, even when given after the dust exposure (49); initial therapeutic trials in humans are under way. In our own laboratory, we have confirmed that the hemolytic potency of chrysotile in vitro can be altered or eliminated by appropriate surface treatments (50). Should in vivo studies suggest that the biological effect of chrysotile is similarly altered, we could conceive of a "safe chrysotile."

Non-fibrotic Lung Disease

Classical stigmata of occupational lung disease have included the nodular fibrosis of silicosis and the diffuse interstitial fibrosis of asbestosis, hard metal disease, and aluminosis. New variants have been added with the fibrosis of extrinsic allergic alveolitis, and the granulomatous lesions of beryllium lung disease.

It is now recognized that fibrosis is not the only possible development. Gough has described the pulmonary rheumatoid nodules of Caplan's Syndrome complicating coal workers' pneumoconiosis (51), and pleural plaques and pleural calcification are recognized as prominent consequences of asbestos inhalation (52,53). Neoplastic changes in a variety of occupational lung diseases include both bronchogenic carcinoma and pleural mesothelioma. The question has even been raised as to whether changes usually associated with some "collagen" diseases might not also be associated with coal workers' pneumoconiosis (54), and silica exposure has been implicated as a cause of alveolar proteinosis (55). The relation between pulmonary emphysema and occupational exposure to dusts, fumes, and vapors has been incompletely studied, but some epidemiological data are available to suggest that they may be of considerable importance in addition to, and particularly when associated with, cigarette smoking (56).

Non-pulmonary Effects

It has become evident that biologically active agents, once inhaled, may affect much more than the lung.

Extrapulmonary dissemination has long been known; silica particles are easily found in the spleen and investigations have recently demonstrated wide dissemination of asbestos fibers in both human (57) and animal subjects (58). Dissemination may be through the lymphatics and blood vessels or by swallowing of particles brought up from the lung by ciliary action. The exact mechanisms are not always clear; the route of contamination which results in peritoneal mesothelioma among asbestos workers, for example, has not been definitely determined. It may follow the ingestion of fibers, with transmural migration through the bowel wall, dissemination by blood vessels to the peritoneum, or even retrograde transdiaphragmatic migration via lymphatics.

Systemic reactions are evidenced by serological changes. In addition to Caplan's Syndrome, a rheumatoid factor is frequently found in the serum of coal workers and asbestos workers without rheumatoid arthritis (59). Antibodies are found in the blood of individuals exposed to detergent enzymes (60). Serum protein changes are found in a variety of other occupational lung diseases.

ENVIRONMENTAL CONSIDERATIONS

Evolution Theory

The long lapse period between exposure and disease makes it difficult to judge whether a new material or a new exposure will or will not be hazardous. Prediction is difficult or impossible. A hypothesis can be offered as a background for our thinking on these matters: that those exposures which have not been part of man's environment for very long periods of time, and have been but recently added, are suspect. As our knowledge of effects of occupational exposures to dusts, fumes, and vapors broadens, it may be possible to test this hypothesis by evaluation of disease associated with exposure to other agents.

Epidemiological Caveats

In addition to the problems of a long lapse period and the multiple factor effect, there are a number of other epidemiological constraints of importance. First, while occupational lung diseases may be specific, they are not unique, so that it is necessary to define "background" disease which might occur in the same population. This is not always easy, especially when the reaction in question does not occur in very great excess among the employed groups. Another important problem is the selection of controls. Other employed groups that look like good controls may have exposures of their own which may give effects in the same direction. "Non-exposed" individuals in the same factory or industry may have low level or intermittent exposure to the agent being studied. Indirect occupational exposure is common in such industries as the construction trades, shipyard work, steel making, farming, etc.

Estimating the environmental hazard to the community at large from exposure to agents which cause occupational lung disease, is beset with all these difficulties (61). There is the further problem of identifying occupationally exposed individuals who are also part of the "general population" (62).

REFERENCES

1. Ramazzini, B. (1713). De Morbis Artificum (Diseases of Workers). Translated by Wilmer C. Wright. Hafner, New York, 1964.

2. Agricola, G. (1556). De Re Metallica. Translated by Herbert C. Hoover and Lou H. Hoover. Dover Publications, New York, 1950.

3. Rosen, G. (1943). The History of Miners' Diseases. Schuman's, New York.

4. Meiklejohn, A. (1951,1952). History of lung diseases of coal miners in Great Britain: Part I. 1800-1875. Brit. J. Ind. Med. 8: 127. Part II. 1875-1920. Brit. J. Ind. Med. 9: 93. . Part III. 1920-1952. Brit. J. Ind. Med. 9: 208.

5. Collis, E. L. (1915). Milroy Lecture. Industrial pneumoconiosis, with special reference to dust-phthisis. Public Health (London), 28: 252, 292; 29: 11, 37.

6. Nagelschmidt, G. (1960). The relation between lung dust and lung pathology in pneumoconiosis. Brit. J. Ind. Med. 17: 247.

7. Selikoff, I. J., Churg, J. and Hammond, E. C. (1964). Asbestos exposure and neoplasia. J. Amer. Med. Assoc. 188: 22.

8. Selikoff, I. J., Hammond, E. C. and Churg, J. (1968). Asbestos exposure, smoking, and neoplasia. J. Amer. Med. Assoc. 204: 106.

9. Selikoff, I. J., Churg, J. and Hammond, E. C. (1965). The occurrence of asbestosis among insulation workers in the United States. Ann. New York Acad. Sci. 132: 139.

10. Siegal, W., Smith, A. R. and Greenburg, L. (1943). Study of talc miners and millers in St. Lawrence County, New Jersey. Ind. Bull. 22: 434, 468.

11. Kleinfeld, M. et al. (1967). Mortality among talc miners. Arch. Env. Health 14: 663.

12. Heimann, H. et al. (1953). Silicosis in mica mining in Bihar, India. Arch. Ind. Hyg. and Occup. Med. 8: 420.

13. Edenfield, R. W. (1960). A clinical and roentgenological study of kaolin workers. Arch. Env. Health 1: 392.

14. Barrie, H. J. and Gosselin, L. (1960). Massive pneumoconiosis from rock dust containing no free silica; nephaline lung. Arch. Env. Health 1: 109.

15. Key, M. M., Kerr, L. and Bundy, M. (1971). Pulmonary Reactions to Coal Dust. Academic Press, New York.

16. Collis, E. L. and Gilchrist, J. C. (1928). Effects of dust upon coal trimmers. J. Ind. Hyg. 10: 101.

17. Gilson, J. C. (1957). Pathology and epidemiology of coal workers, pneumoconiosis in Wales. Arch. Ind. Health 15: 468.

18. Rogan, J. M., Rae, S. and Walton, W. H. (1965). The National Coal Board's Pneumoconiosis Field Research -- An interim review. In: C. M. Davies, (ed.). Inhaled Particles and Vapours II. Pergamon Press, London, 493.

19. McLaughlin, A. I. G. (1950). Industrial lung disease of iron and steel foundry workers. Her Majesty's Stationery Office, London.

20. Robertson, A. J. (1964). The romance of tin. Lancet I: 1229, 1289.

21. Browning, E. (1961). Toxicity of Industrial Metals. Butterworth, London.

22. Hardy, H. L. (1946). Delayed chemical pneumonitis occurring in workers exposed to beryllium compound. J. Ind. Hyg. 28: 197.

23. Goralewski, G. (1941). Zur klinik der aluminiumlunge. Arch. Gewerbepath. 11: 106.

24. Koven, A. L. et al. (1953). Health of workers in the chromate producing industry. A study. PHS Publication No. 192, U. S. Govt. Printing Office.

25. Sunderman, F. W. (1968). Nickel carcinogenesis; epidemiology of respiratory cancer among nickel workers. Dis. Chest 54: 41.

26. Shaver, C. G. and Riddel, A. R. (1947). Lung changes associated with manufacture of alumina abrasives. J. Ind. Hyg. and Toxicol. 29: 145.

27. Annotation. Hard-metal disease. (1963). Brit. Med. J. I: 836.

28. Langer, A. M., Selikoff, I. J. and Sastre, A. (1971). Chrysotile asbestos in the lungs of persons in New York City. Arch. Env. Health 22: 348.

29. Boren, H. G. (1967). Pathobiology of air pollutants. Env. Res. 1: 178.

30. Eto, H. (1961). Studies on pneumoconiosis by the inhalation of amorphous carbon. J. Nara Med. Assoc. 12: 411.

31. Pendergrass, E. P. et al. (1967,1968). Observations on workers in the graphite industry. Med. Radiogr. and Photogr. Part 1. 43: 70 . Part 2. 44: 2 .

32. Section of Occupational Medicine, Roy. Soc. Med. (1969). Symposium on Isocyanates. Proc. Roy. Soc. Med. 63: 365.

33. Flindt, M. L. H. (1969). Pulmonary disease due to inhalation of derivatives of Bacillus subtilis containing proteolytic enzyme. Lancet I: 1188.

34. Lloyd, J. W. (1971). Long-term mortality study of steel workers. V. Respiratory cancer in coke plant workers. J. Occup. Med. 13: 53.

35. Bouhuys, A. et al. (1967). Byssinosis in the United States. New Eng. J. Med. 277: 170.

36. Dickie, H. A. and Rankin, J. (1958). Farmer's lung. An acute granulomatous interstitial pneumonitis occurring in agricultural workers. J. Amer. Med. Assoc. 167: 1069.

37. Newhouse, M. L., Tagg, B. and Pocock, S. J. (1970). An epidemiological study of workers producing enzyme washing powders. Lancet I: 689.

38. Buechner, H. A. (1961). The industrial compensability of bagassosis. Ind. Med. Surg. 30: 294.

39. Pepys, J. (1969). Hypersensitivity Diseases of the Lungs due to Fungi and Organic Dusts. Monographs in Allergy, Vol. 4, Karger, Basel, Switzerland.

40. McCallum, R. I. (1968). Decompression sickness. Brit. J. Ind. Med. 25: 4.

41. Pirchan, A. and Sikl, H. (1932). Cancer of the lung in the miners of Jachymov (Joachimstal). Report of cases observed in 1929-1930. Amer. J. Cancer 16: 681.

42. de Villiers, A. J. and Windish, J. P. (1964). Lung cancer in a fluorspar mining community. I. Radiation, dust, and mortality experience. Brit. J. Ind. Med. 21: 94.

43. Policard, A. (1962). The conflict of living matter with the mineral world: the pneumoconioses. J. Clin. Path. 15: 394.

44. Sluis-Cremer, G. K., Walters, L. G. and Sichel, H. S. (1967). Chronic bronchitis in miners and non-miners: An epidemiological survey of a community in the gold-mining area in the Transvaal. Brit. J. Ind. Med. 24: 1.

45. Lundin, F. E. et al. (1969). Mortality of uranium miners in relation to radiation exposure, hard-rock mining and cigarette smoking - 1950 through September 1967. Health Phys. 16: 571.

46. Selikoff, I. J. and Hammond, E. C. (1971). Lapsed Period. Env. Res. 4(2): ii.

47. Turkevich, J. (1959). The world of fine particles. Amer. Sci. 47: 97.

48. Langer, A. M. et al. (In press). Inorganic fibers, including chrysotile, in lungs at autopsy; preliminary report. Proceedings British Occupational Hygiene Society - Third International Symposium on Inhaled Fibers, September 1970.

49. Schlipkoter, H. W. (1967). Progress in silicosis research. Proc. 5th International Silicosis Conference, Munster. Ind. Hyg. Dig. 2: 189.

50. Schnitzer, R. J. and Pundsack, F. L. (1970). Asbestos hemolysis. Env. Res. 3: 1.

51. Gough, J., Rivers, D. and Seal, R. M. E. (1955). Pathological studies of modified pneumoconiosis in coalminers with rheumatoid arthritis (Caplan's Syndrome). Thorax 10: 9.

52. Kiviluoto, R. (1960). Pleural calcification as a roentgenologic sign of non-occupational endemic anthrophyllite-asbestosis. Acta Radiol. (Suppl.) 194: 1.

53. Selikoff, I. J. (1965). The occurrence of pleural calcification among asbestos insulation workers. Ann. New York Acad. Sci. 132: 351.

54. Rodnan, G. et al. (1967). The association of progressive systemic sclerosis (scleroderma) with coal miners' pneumoconiosis and other forms of silicosis. Ann. Int. Med. 66: 323.

55. Buechner, H. A. and Ansari, A. (1969). Acute silico-proteinosis. Dis. Chest 55: 274.

56. Hammond, E. C. and Selikoff, I. J. (1970). The effects of air pollution; epidemiological evidence. In: H. A. Shapiro, (ed.). Proceedings International Conference on Pneumoconiosis, Johannesburg, 1969. Oxford University Press, Capetown, 368.

57. Selikoff, I. J. (1965). Discussion. Ann. New York Acad. Sci. 132: 596.

58. Kanazawa, K. et al. (1970). Migration of asbestos fibres from subcutaneous injection sites in mice. Brit. J. Cancer, 24: 96.

59. Turner-Warwick, M. and Parkes, W. R. (1970). Circulating rheumatoid and antinuclear factors in asbestos workers. Brit. Med. J. II: 492.

60. Berson, S. A. et al. (1971). Antibodies to "alcalase" after industrial exposure. New Eng. J. Med. 284: 688.

61. Committee on Biological Effects of Atmospheric Pollutants. (1971). Asbestos -- The need for and feasibility of air pollution controls. Nat. Acad. Sci., Washington, D. C.

62. Gilson, J. C. (1970). Occupational Bronchitis (?). Proc. Roy. Soc. Med. 63: 857.

CHAPTER 14. EPIDEMIC ASTHMA

ROBERT J. M. HORTON, National Environmental
Research Center, Environmental Protection Agency,
Research Triangle Park, North Carolina

Acute manifestations of atopic allergy, e.g.
attacks of bronchial asthma or of allergic rhinitis,
do not usually occur in epidemics. Allergic
individuals are sensitive to a great variety of
substances in their environments, and are disturbed
by a variety of irritants also. Under ordinary
circumstances allergic attacks occurring in
populations are scattered in time and place. But a
few instances have been identified and studied in
which epidemics of asthmatic attacks have been
observed to occur repeatedly, for short times, in the
same place. These situations will be discussed
below, particularly the studies made in New Orleans
and Minneapolis. On occasion a single dramatic
episode is reported (1). Ill defined sharp increases
in asthma have also been reported to occur during the
famous severe air pollution episodes.

SEASONAL VARIATIONS

Efforts have been made to find epidemic patterns
of asthma in some localities (2-5). These have
mainly delineated the normal pattern of asthmatic
attack treatments to be found by examining the
records of emergency facilities of community
hospitals, where the level of attacks treated shows
modest oscillations from day to day in most
instances, but otherwise remains quite uniform
throughout most of the year. In all of the cases
examined there is a single season of increase from
mid-September to mid-December. The same general
pattern has been observed in the records of
treatments by private practitioners (6). This autumn
increase has also been frequently commented upon by

allergists in relation to their practices. A number of possible explanations has been offered for this seasonal rise. None have been well established. Spore and pollen allergens with seasons of prevalence at this time have been suggested but not substantiated. Resuspension of settled allergens by the institution of heating in homes has been suggested but not carefully evaluated. The seasonal pattern of prevalence of house dust mites offers a new interesting possibility (7). Seasonal acclimatization to cold weather is also a suggestion which merits careful consideration. Allergic individuals are frequently considered to be less able than the general population to adjust to cold weather. The observations of Greenburg, et al. in New York City on the effects of extreme cold waves early in the fall are of particular interest in this regard (4-5). They found in analyzing the emergency room records of asthmatic treatments of several hospitals that in some years sharp brief increases occurred in all the hospitals simultaneously. These epidemic episodes coincided with very severe early cold spells. They were not observed to accompany severe cold waves which occurred later in the year. This type of episode, and its relationship to the general autumn increase in asthma attacks merits further study.

It should be noted here that the above studies do not show any seasonal increase in asthmatic attacks during the pollen seasons commonly associated with increases in seasonal allergic rhinitis (hay fever, pollenosis). Asthma does occur frequently with the rhinitis. Diary records of asthmatic attacks kept by patients do show appropriate seasonal increases in spring, early summer, and later summer. They also show the autumn rise described above. The explanation for this difference appears to lie in the observation that asthmatic attacks accompanying pollen allergy are milder than others, are self-treated and therefore do not appear in the records of emergency medical facilities.

CASTOR POMACE ALLERGY

The best known and most thoroughly studied form of epidemic asthma is that due to castor pomace. This substance is the fine powder which remains after removing castor oil from the castor bean. The extensive literature on this subject has been recently reviewed and a complete bibliography furnished (8). Castor pomace is one of the most potent allergens known. It has long been recognized for its troublesome characteristics in industry. Most firms producing or handling this substance are careful not to hire allergic individuals. Since the pomace is so fine and light, especially when expression of oil is followed by solvent extraction, it is readily dispersed into the ambient air surrounding industrial plants which make or use it, and also during transportation. For this reason allergic people living or working near such sources become sensitized to the pomace and subject to attacks on subsequent exposure, unless very good control methods are applied to restrict the dispersion of the dust. Without these appropriate restrictions, repeated epidemics of severe asthmatic attacks are observed in the neighborhoods about such sources. These occur when suitable plant operations and appropriate meteorological conditions coincide. Cases are observed down wind from the source for several blocks, or, with a large source, for distances in excess of a mile. Fortunately the sources of castor pomace are not numerous. They consist mainly of castor oil mills, and fertilizer factories which use most of the pomace produced in preparing mixed plant foods. Modern industrial hygiene techniques are capable of greatly reducing the spread of the dust when they are properly applied.

NEW ORLEANS ASTHMA

Epidemics of bronchial asthma in the central area of the city of New Orleans have been found to occur since 1953 at least, and have been extensively studied for over a decade (9-23). Most of the earlier work has been reviewed in the two most recent reports (22,23). The epidemics are apparent in the emergency room records of the Charity Hospital, daily treatments for this cause varying from 15 to a maximum of 250. Such extreme fluctuations have not been observed elsewhere. Brief outbreaks occur regularly in the fall, and sporadically at other times of the year. They are characteristically found during periods of very low wind velocity, air stagnation, low humidity. Such periods frequently occur following a decline in temperature. Such meteorological conditions favor the accumulation of locally produced air pollutants. New Orleans does not have very high levels of conventional pollutants, e.g. sulfur oxides, oxidants. It does have an unusual number of sources of pollution from organic dusts such as coffee, grain, and bagasse. It has been thought that one or more of such sources would be found responsible for the observed outbreaks. However none has been clearly demonstrated to be responsible in whole or in part, nor have any been clearly exonerated. The illness found in the outbreaks is quite clearly of an allergic nature. It is not unusual or unique. There is no separate group of asthmatics who come in for treatment only during outbreaks. The prevalence of active asthmatics in the central city is not higher than that found elsewhere. The outbreaks are essentially limited to the adult population. No consistent relationship between any of numerous particulate air pollutants measured and the asthma epidemics has been observed. One is left with the impression that the outbreaks are due a combination of one or more local allergens of uncertain identity with appropriate meteorological conditions.

MINNEAPOLIS ASTHMA

The presence of epidemic asthma associated with the grain industry in Minneapolis was first suggested by Wittich (24) from observations on his patients. More recently the part of this problem found in the student body at the University of Minnesota has been the subject of intensive investigation (25-34). The most recent of these reports (34) reviews much of the previous work. The University's campus in Minneapolis is adjacent to a large industrial complex consisting mainly of grain storage, transport, and use in the production of flour, cereals, and other products. There is considerable air pollution with organic dusts from these sources. Some pollution with sulfur and nitrogen oxides and undifferentiated particulates from power generation and heating is also present. Epidemics of attacks of asthma among allergic students were observed to occur particularly in conjunction with periods of increased air pollution. Intensive study of student attacks, and of air pollution and meteorologic parameters over a number of years, was carried out. Sophisticated chemical identification and separation of protein allergens from collected particulate matter from air permitted specific skin testing of patients for sensitivity. Positive relationships were shown between student asthmatic attacks and air pollution with both particles and sulfur dioxide. The relation between particulate pollution and asthmatic attacks was closer for those students who reacted positively to skin testing with the specific allergens referred to above than for those not so reacting. Positive relationship of asthma to humidity, and negative correlation with temperature were also observed, but these were weaker than the associations with the air pollution parameters. It is apparent here that multiple environmental factors influence the presence of attacks of asthma in the students under observation. The strongest factors, especially in those specifically sensitive thereto, are specific airborne allergens from the adjacent grain industry complex, and an irritant gaseous air pollutant sulfur dioxide.

YOKOHAMA ASTHMA

This respiratory illness was first observed in U. S. military personnel who were stationed in and near the city of Yokohama, Japan in 1949. Continuing observations and studies resulted in an initial report by Huber et al. in 1954 (35). Further investigations by several military physicians have been published since (36-46). Early observations brought out the similarities of the Yokohama phenomena to epidemic asthma. Continued study made it apparent that the illnesses being observed were not asthma, nor were they allergic. The clinical picture is that of a chronic bronchitis with wheezing, sometimes called asthmatic bronchitis. Epidemics of wheezing attacks were produced in affected persons by high levels of air pollution. It is mentioned here only to indicate that the name is misleading, and that this is not an example of epidemic asthma.

SUMMARY

Although bronchial asthma is not usually an epidemic disease, situations do occur in which numerous asthmatics have attacks simultaneously or nearly so in a limited area and time. The most conspicuous and well defined of such phenomena have been found to be associated with repeated exposure to air pollution by allergenic vegetable dusts from industrial sources. Other causes of episodes have been air pollution with irritant gases and sharp decreases in temperature. Combinations of all or several of these stresses may also be responsible. Outbreaks of wheezing in asthmatic bronchitis may also be induced by increased air pollution; they resemble epidemic asthma in some respects.

REFERENCES

1. Morrison, I. and Bath, A. T. (1960). It happened one night. Med. J. Australia 1: 850.

2. Booth, S. et al. (1965). Detection of asthma epidemics in seven cities. Arch. Env. Health 10: 152.

3. Newill, G. R. (1963). Predicting asthma outbreaks. Pub. Health Rpts. 78: 259.

4. Greenburg, L. et al. (1964). Asthma and temperature change. Arch. Env. Health 11: 642.

5. Greenburg, L. et al. (1967). Asthma and temperature change. In: S. W. Tromp and W. H. Weihe, (eds.). Biometeor. 2: 3, Pergamon Press, New York.

6. National Disease and Therapuetic Index. (1964). Spec. Med. Rpt., Seasonal Patterns of Asthma Attacks. Lea Associates, Ambler, Pennsylvania.

7. Voorhorst, R. et al (1967). The house dust mite D. pteronyssinus and the allergens it produces. J. Allergy 39: 325.

8. Apen, E. M. et al. (1967). Health Aspects of Castor Bean Dust. Pub. Health Serv. Pub. No. 999-AP-36, Washington.

9. Lewis, R. F. and Cleve, F. A. (1960). Meteorologic aspects of New Orleans asthma. Air Pollution Medical Program, U.S. Public Health Service, Washington.

10. Lewis, R. F. et al (1962). Air Pollution and New Orleans Asthma, 2 vols., Tulane Univ., New Orleans.

11. Lewis, R. F. et al. (1962). Air Pollution and New Orleans Asthma. Pub. Health Rpts. 77: 947.

12. Newell, G. R. and Swafford, L. I. (1963). Epidemiology of asthma in children with particular reference to wind speed and wind direction. Pediat. 31: 134.

13. Lewis, R. F. (1963). Epidemic asthma in New Orleans, a summary of knowledge to date. J. Louisiana State Med. Soc. 115: 300.

14. Weill, H. et al. (1964). Further observations on New Orleans asthma. Arch. Env. Health 8: 184.

15. Weill, H. et al. (1964). Epidemic asthma in New Orleans. J. Amer. Med. Assoc. 190: 811.

16. Walsh, J. J. and Derbes, V. J. (1964). The fifth season, New Orleans asthma. Bull. Nat. Tbc. Assoc., (Dec.), 8.

17. Weill, H. et al. (1965). Recent developments in New Orleans asthma. Arch. Env. Health 10: 148.

18. Weill, H. et al. (1965). Allergenic air pollutants in New Orleans. J. Air Poll. Cont. Assoc. 15: 467.

19. DeMarrais, G. A. (1965). The meteorology associated with New Orleans asthma. Arch. Env. Health 11: 787.

20. Kenline, P. A. (1966). October 1963 New Orleans asthma study. Arch. Env. Health 12: 295.

21. Salvaggio, J. E. and Klein, R. C. (1967). New Orleans asthma. I. Characterization of individuals involved in epidemics. J. Allergy 39: 227.

22. Carroll, R. E. (1968). Epidemiology of New Orleans epidemic asthma. Amer. J. Pub. Health 58: 1677.

23. Salvaggio, J. et al. (1970). New Orleans asthma. II. Relationship of climatologic and seasonal factors to outbreaks. J. Allergy 45: 257.

24. Wittich, F. W. (1952). Respiratory tract allergic effects from chemical air pollution. In: L. C. McCabe, (ed.). Air Pollution. McGraw Hill Book Company, New York.

25. McLouth, M. E. and Paulus, H. J. (1961). Air pollution from the grain industry. J. Air Poll. Cont. Assoc. 11: 313.

26. Goppers, V. and Paulus, H. J. (1961). Isolation of proteins from samples of air-borne particulates. Am. Ind. Hyg. Assoc. J. 22: 54.

27. Goppers, V. and Paulus, H. J. (1962). Macromolecular compounds isolated from air-borne particles by electrophoresis and paper chromatography. Am. Ind. Hyg. Assoc. J. 23: 181.

28. Cowan, D. W. et al. (1963). Bronchial asthma associated with air pollutants from the grain industry. J. Air Poll. Cont. Assoc. 13: 546.

29. Mill, R. A. et al. (1965). Measuring the environment for a bronchial asthma study. Am. Ind. Hyg. Assoc. J. 26: 510.

30. Goppers, V. and Paulus, H. J. (1965). Preparative separation of allergenic macromolecular compounds from airborne particles using continuous electrophoresis. J. Chromat. 17: 628.

31. Goppers, V. and Paulus, H. J. (1966). Purification of allergenic macromolecular compounds isolated from airborne particles on thin-layer and preparative chromatography. Am. Ind. Hyg. Assoc. J. 27: 144.

32. Paulus, H. J. and Smith, T. S. (1967). Association of allergic bronchial asthma with certain air pollutants and weather parameters. Int. J. Biometeor. 11: 119.

33. Mill, R. A. (1967). Air Pollution and the Grain Industry. Doct. Thesis, Univ. Minnesota, Minneapolis.

34. Goppers, V. and Paulus, H. J. (1967). Allergenic compounds in nature. Internat. Arch. Allergy Appl. Immunol. 31: 546.

35. Cowan, D. W. and Paulus, H. J. (1969). Relationship of air pollution to allergic diseases. Final Rpt. PHS Res. Grant AP-00090, Univ. Minnesota, Minneapolis.

36. Huber, T. E. et al. (1954). New environmental respiratory disease (Yokohama asthma). Arch. Ind. Hyg. Occup. Med. 10: 399.

37. Phelps, H. W. et al. (1961). Air pollution asthma among military personnel in Japan. J. Amer. Med. Assoc. 175: 990.

38. Phelps, H. W. (1961). Air pollution asthmatic bronchitis among United States personnel in Japan. Jap. Heart J. 2: 180.

39. Phelps, H. W. (1961). Pulmonary function studies used to evaluate air pollution asthma disability. Milit. Med. 126: 282.

40. Phelps, H. W. and Koike, S. (1962). Tokyo - Yokohama asthma. Amer. Rev. Resp. Dis 86: 55.

41. Ladd, B. and Phelps, H. W. (1963). Incidence of air pollution bronchitis in military personnel in Japan. Dis. Chest 43: 151.

42. Smith, R. B. et al. (1964). Tokyo - Yokohama asthma, an area specific disease. Arch. Env. Health 8: 805.

43. Oshima, Y. et al. (1964). A study of Tokyo-Yokohama asthma among Japanese. Amer. Rev. Resp. Dis. 90: 632.

44. Beard, R. et al. (1964). Observations on Tokyo - Yokohama asthma and air pollution in Japan. Pub. Health Rpts. 79: 439.

45. Phelps, H. W. (1965). Follow up of cases of Tokyo-Yokohama respiratory disease. Arch. Env. Health 10: 143.

46. Spotnitz, M. (1965). The significance of Yokohama asthma. Amer. Rev. Resp. Dis. 92: 371.

47. Spotnitz, M. (1966). Air pollution respiratory disease. A military medical problem. Milit. Med. 131: 1499.

CHAPTER 15. ENVIRONMENTAL FACTORS IN BRONCHIAL ASTHMA

CARL M. SHY, VICTOR HASSELBLAD,
LEO T. HEIDERSCHEIT, and ARLAN A. COHEN
National Environmental Research Center
Environmental Protection Agency
Research Triangle Park, North Carolina

The frequency of asthma in a community may be influenced by four environmental variables: concentrations of aeroallergens, season of year, climatic conditions, and levels of air pollutants. The sensitivity of asthmatics to pollens, spores, molds, and other aeroallergens varies widely from individual to individual and has been a subject of much clinical investigation by allergists. This environmental factor will not be considered in the present discussion. (See Chapter 10.)

SEASON OF THE YEAR

The peak incidence of asthma has been repeatedly observed during the months of October and November (1-3) in the northern hemisphere and during April and May (4,5) in the southern hemisphere. The percent distribution of asthma for each month of 1960 in four large metropolitan hospitals of the United States (Table 1) illustrates this seasonal phenomenon. No one environmental factor explains the autumnal peak in asthma incidence, although several possibilities have been considered, including onset of the heating season (2) and passage of cold fronts (4,6).

TABLE 1 -- PERCENT DISTRIBUTION OF ASTHMA BY MONTH AT
FOUR LARGE METROPOLITAN HOSPITALS, 1960 (1)

	Harlem N. Y.	Metropolitan N. Y.	Charity New Orleans	Cook County Ill.
Jan.	8.48	7.49	5.94	5.96
Feb.	7.59	7.33	5.12	4.93
Mar.	7.69	7.93	4.08	4.64
Apr.	5.75	5.96	5.22	5.53
May	6.23	6.55	8.18	7.09
Jun.	5.82	5.95	5.89	7.72
Jul.	5.02	5.58	8.00	6.30
Aug.	7.08	5.72	6.64	7.46
Sep.	8.10	7.47	7.93	10.30
Oct.	13.04	14.76	15.01	17.34
Nov.	13.63	14.41	17.10	12.78
Dec.	11.56	10.82	10.89	9.96

Large and episodic increases in asthma frequency have been superimposed on the autumnal peak in New Orleans since 1953 (7-9). Burning dumps (7), grain dusts, and meteorologic variables (8) have been implicated, but no satisfactory explanation for individual episodes was ever found. Recently, Salvaggio observed the association of the New Orleans asthma episodes with high mold-spore counts on days of low wind and low humidity in fall (10). He hypothesized that the many bayous and swamps in close proximity to New Orleans favored mold growth and that stagnant air favored concentration of these molds in the atmosphere.

CLIMATIC CONDITIONS

Cold temperatures or temperature drops have a strong influence on asthma frequency. Hasselblad found statistically significant associations between New Orleans asthma episodes and days with temperature below 45°F in October and November, 1963 through 1968 (11). For eight months Cohen followed daily variations in the asthma status of 20 subjects living near a large source of sulfur dioxide and particulate pollution (12). As shown in Table 2, attack rates were more highly correlated with temperature than with pollutant levels or other meteorologic variables. The negative correlation coefficient for temperature signifies increasing attack rates with declining temperature. Derrick (4) and Tromp (6) associated marked increases in asthma with the passage of cold fronts, and Greenburg related onset of the heating season in fall to notable increases in asthma frequency (13).

Other climatic conditions have been less consistently related to asthma. Girsh and associates (3) in Philadelphia reported that asthma attacks in childhood were significantly more frequent on days of high barometric pressure. However, the authors failed to isolate seasonal and temperature effects on asthma frequency, and the distinct possibility of confounding barometric pressure with season and temperature was not ruled out.

AIR POLLUTION

Asthmatic patients are particularly vulnerable to high levels of atmospheric pollution. Schrenk and colleagues (14) made an intensive study of the Donora population after that community's acute air pollution episode of October 1948. Ninety percent of the asthmatic population, compared with 40 percent of the total population, experienced acute symptoms during the episode.

Several prospective surveys have shown relationships between asthma attack rates and levels of atmospheric pollutants. Zeidberg found an association of asthma attacks in adults, but not children, with levels of sulfation in Nashville (15). A panel of asthmatics in Los Angeles had significantly more attacks on days when hourly

TABLE 2 -- CORRELATION COEFFICIENTS FOR ASTHMA ATTACK
RATE WITH POLLUTION AND CLIMATIC VARIABLES (12)

Variable Correlated With Attack Rate	Correlation Coefficient	p Value
Temperature	-.427	<.01
Soiling index	+.387	<.01
SO_2	+.320	<.01
Total particulates	+.241	<.01
Suspended sulfates	+.199	<.01
Suspended nitrates	+.169	<.01
Barometric pressure	+.038	NS
Wind speed	+.050	NS
Humidity	+.112	NS

photochemical oxidant levels exceeded 0.13 ppm - a commonly exceeded threshold in Los Angeles during the summer and fall oxidant season (16). Philadelphia children were reported to experience significant increases in asthma attack rates on days when pollutant concentrations were in the upper decile in frequency distribution (3), but possible confounding effects with temperature and season were not analyzed.

The most quantitative information on dose levels of pollutants in relation to asthma frequency was reported in the study by Cohen and associates (12). As noted previously, declining temperature had the strongest effect on attack rates, but even after this variable was isolated, a significant effect of atmospheric pollutants could be shown (Table 3). Since all measured pollutants originated from the same point source -- a large coal fueled power plant -- it was not possible for the authors to disentangle the influence of individual pollutants on attack rates. However, as shown in Figure 1, the strength of pollutant exposure on asthma frequency varied within different temperature ranges. At low temperatures in the range of 0 to 30°F, increasing levels of SO_2 had relatively less influence on attack rates than when temperatures were 31 to 50°F or 50°F and above. These data suggest that cold temperature obtains a nearly maximal response from susceptible asthmatics, and that air pollutants exert their most

TABLE 3 -- RESIDUAL VARIATION IN ASTHMA ATTACKS RATE
EXPLAINED BY EACH VARIABLE AFTER EFFECTS
OF TEMPERATURE ARE REMOVED (12)

Factor	Percent of Variance Explained	p Value
Temperature alone	20.3	.00005
Temperature with:		
Soiling	3.92	.006
SO_2	3.95	.006
Particulates	3.86	.006
Nitrates	2.36	.033
Sulfates	5.48	.001
Barom. Press.	0.00	NS
Wind Speed	0.17	NS
Humidity	0.18	NS

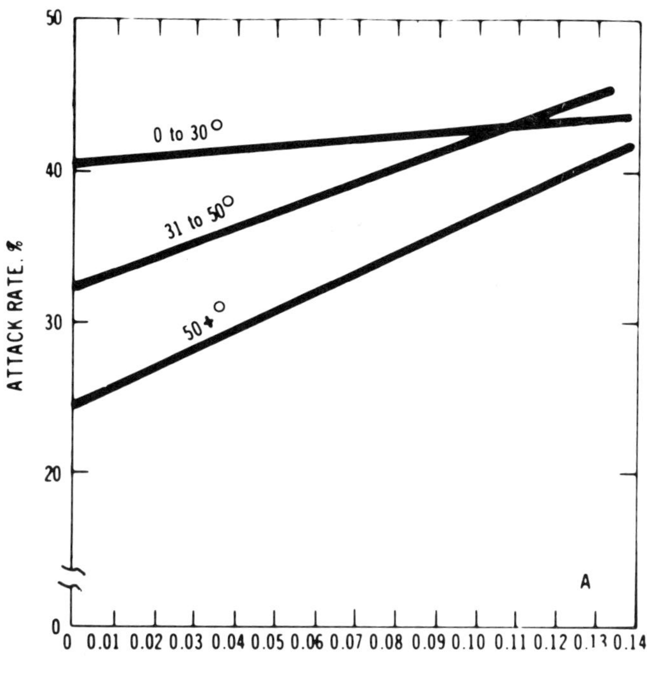

Fig. 1. Asthma Attack Rate vs. Daily Sulfur Dioxide Concentrations within Three Temperature Ranges (12).

deleterious, independent, asthmatogenic effect when temperatures are more moderate. Almost identical pollutant-temperature relationships with attack rates were found for each of the pollutants listed in Table 3.

Ishizaki recently reported a study in which asthmatics and normal subjects were experimentally exposed to sulfur dioxide over the range of 0 to 25ppm (17). $FEV_{1.0}$, airway resistance and changes in functional residual capacity were exquisitely sensitive to increasing sulfur dioxide among asthmatics, but not among the normal subjects. The adverse effect of SO_2 could be completely prevented by administration of atropine sulfate, suggesting mediation of the SO_2 effect through the parasympathetic nervous system.

SUMMARY AND CONCLUSIONS

Three environmental factors influence asthma rates in a community: the fall season, cold temperatures, and air pollution. These determinants of asthma frequency should be considered in the design and analysis of studies attempting to elucidate etiological or contributing factors in asthma attacks. Cold temperatures appear to exert a stronger effect than atmospheric pollutants on attack rates. At moderate temperature ranges (50°F and above) air pollutants have a relatively stronger independent effect on asthma frequency, and this effect was observed at atmospheric levels commonly experienced in urban areas.

REFERENCES

1. Booth, S. et al. (1965). Detection of asthma epidemics in seven cities. Arch. Env. Health 10: 152.

2. Greenberg, L. and Field, F. (1965). Air pollution and asthma. J. Asthma Res. 2: 195.

3. Girsh, L. S. et al. (1967). A study on the epidemiology of asthma in children in Philadelphia. J. Allergy 39: 347.

4. Derrick, E. H. (1966). The annual variation of asthma in Brisbane: Its relation to the weather. Int. J. Biometeor. 10: 91.

5. Derrick, E. H., Thatcher, R. H. and Trappett, L. G. (1960). The seasonal distribution of hospital admissions for asthma in Brisbane. Australian Ann. Med. 9: 180.

6. Tromp, S. W. (1962). Biometeorological analysis of the frequency and degree of asthma attacks in the western part of the Netherlands. In: S. W. Tromp (ed.), Proceedings of the Second International Bioclimatological Conference. Pergamon Press, Oxford, 477.

7. Lewis, R., Gilkeson, M. M. and McCaldin, R. O. (1962). Air pollution and New Orleans asthma. Pub. Health Rep. 77: 947.

8. Weill, H. et al. (1964). Epidemic asthma in New Orleans. J. Amer. Med. Assoc. 190: 811.

9. Carroll, R. E. (1968). Epidemiology of New Orleans epidemic asthma. Amer. J. Pub. Health 58: 1677.

10. Salvaggio, J. and Seabury, J. (1971). New Orleans asthma. 4. Semiquantitative airborne spore sampling, 1967 and 1968. J. Allergy and Clin. Immunol. 48: 82.

11. Hasselblad, V. and Heiderscheit, L. T. (1970). Summary of information on asthma in New Orleans. In-house technical report, Division of Health Effects Research, Environmental Protection Agency.

12. Cohen, A. A. et al. (1971). Asthma and air pollution from a coal-fueled power plant. Amer. J. Pub. Health. (In press).

13. Greenburg, L. et al. (1967). Asthma and temperature change. Biometeor. 2: 3.

14. Schrenk, H. H. et al. (1949). Air pollution in Donora, Pennsylvania: Epidemiology of the unusual smog episode of October 1948. Pub. Health Bull. 306.

15. Zeidberg, L. D., Prindle, R. A. and Landau, E. (1961). The Nashville air pollution study. Amer. Rev. Resp. Dis. 84: 489.

16. Schoettlin, C. E. and Landau, E. (1961). Air pollution and asthmatic attacks in the Los Angeles area. Pub. Health Rpts. 76: 545.

17. Ishizaki, T. (1970). Effect of air pollution upon bronchial asthma. Jap. J. Med. 9: 238.

CHAPTER 16. TOWARDS AN OPTIMUM ENVIRONMENT

ERIC J. CASSELL, Department of Public Health,
Cornell University Medical College, New York

To maintain an optimum environment it is necessary not only to solve present problems, but also to protect against pollutants and sources that are as yet unknown or negligible.

Present trends make it probable that technology will continue to grow and change, power production increase, and transportation remain a problem; in other words, we have not reached a plateau of technological growth that will allow us to consider the environment as a stable though incompletely solved problem.

Generally speaking, laws are drawn to correct things which people have come to think are problems. Because a constituency must be developed among both the society and its legislators that believes that the problem must be solved, the development of legislation is often difficult and time consuming. In the case of air pollution, we have seen how long the process can take. But equally, in the intensity of the present environmental movement, we can see the force that an aroused constituency can provide for the passage of needed laws.

Laws are formed to solve existing problems. But what we know of the problems, such as that of environmental pollution, suggests that problems change. Changing or updating a law is just as difficult as instituting one. It follows that laws based solely on the present common body of beliefs about air pollution stand a good chance of being obsolescent by the time that they are passed.

It is a truism of science that a theory is only as good as its predictive value. It is wise to demand the same requirement of a legislative philosophy. Laws must be drawn on the basis of existing knowledge; but laws can be considered successful only to the extent that they solve problems that arise after they are enacted.

Therefore, a strategy must be developed that provides for the solution of existing problems, but that can also be applied to pollutants and sources of which we are as yet unaware. It is really the pollutants and sources still unknown to us that will provide the test of the adequacy of a control philosophy.

How bound we are by our existing belief structures, and how hard they make it for us to solve new or different problems, deserves some elaboration as it applies to public health.

For the moon landings, elaborate precautions were made to prevent the astronauts from bringing back some new microbe. Or, should they be exposed, from endangering the rest of us. The preparations, precautions, and periods of quarantine were not developed with the knowledge of some specific, infectious agent but rather in the light of a history of successes in the prevention of contagion based on the germ theory of disease. It seems now that these precautions were not necessary -- at least in terms of what we now call infectious agents and disease -- but we knew what to do.

On the other hand, we have been relatively helpless to combat the spread and mortality of cholera in certain economically underdeveloped areas of the world, despite the fact that the cholera vibrio is well known to us, its means of transmission clear, and its contagion definitively halted by simple sanitation. We are stymied in some lesser developed countries because we are unable to get the villager to wash his hands with soap or drink water that has been boiled. The inhabitant resists these simple measures not just out of ignorance, for it has been explained to him that the measures are necessary to kill germs. He resists washing because the whole concept of germ transmission of disease is culturally without meaning. The body of law (in the broadest sense) under which the inhabitant lives and which, in his experience, has successfully dealt with other environmental onslaughts (in the broadest sense), is not "predictive" enough to deal with a new bacterial challenge.

I suggest, though with some timidity, that in relation to the problem of the control of air pollution we may be somewhat in the position of the back country inhabitant.

I would like now to show some evidence that the problem is not fully encompassed by our present understanding; provide a way of looking at it that seems to help; and point to a control philosophy that should help with existing problems and also with those of which are as yet unaware.

Most of us believe that air pollution is a factor in the production, maintenance, or exacerbation of respiratory disease. The evidence has been documented in the preceding chapters. The evidence is of three general types: 1) Toxicological -- man or animals are exposed to single or combined agents under laboratory conditions, and some specific index of health is employed to measure their effects, varying from change in the dynamics of breathing, through the enhancement of infection, to the production of structural change; 2) Physiological -- changes are measured in the normal dynamics of breathing in man or animals exposed to naturally existing conditions of pollution; 3) Epidemiological -- the occurrence of disease or dysfunction is measured in population groups exposed to environmental pollution during the usual activities of life. The methods overlap, but the breakdown is useful for the purpose of understanding the problem.

Toxicologists have shown adverse effects for virtually every identified pollutant. But with few exceptions (e.g., carbon monoxide, ozone, and specific toxicants such as beryllium) the effects are nonspecific and have been shown only at levels that are higher than those found in urban atmospheres. Even some of the exceptions (carbon monoxide and ozone) have not, thus far, been shown to meet their toxicological expectations in urban populations.

Physiologists have repeatedly demonstrated abnormalities in the dynamics of breathing in exposed populations. Using widely varying techniques and widely varying populations (from normal school children to respiratory cripples), adverse effects are seen, but it has not been possible, thus far, to demonstrate what constituent in the atmosphere (including weather) is responsible for the abnormalities. Epidemiologists have also used a great many indices of ill health, from the production of symptoms to the enumeration of deaths, to demonstrate an effect of air pollution on health.

Virtually every index of effect has shown an influence of air pollution; but the specific pollutant responsible for injury has largely escaped conclusive incrimination. For example, at the time of the study many an investigator may have felt that he had indicted a specific pollutant, such as SO_2, but as sophistication has increased, we have become more skeptical of the evidence.

The use of sulfur dioxide as a representative of air pollution exemplifies the difficulties: 1) the techniques of SO_2 measurement generally used rarely measure SO_2 alone, and different techniques of measurement may show considerable disagreement; 2) the presence of SO_2 in the environment is often so closely associated with other pollutants or weather factors that it is impossible to disentangle the effects; 3) the pollutants interact in the environment (and perhaps even in the lung), so that the measurement of a specific pollutant does not adequately measure the dynamic atomsphere; 4) the pollutants, such as SO_2, are closely related to other factors having an effect on health such as poverty, crowding, methods of heating, industrialization, urbanization, and so forth; 5) we know what is in the atmosphere only by what we can measure, so that if there are other unmeasured or unknown components having an effect we cannot know them; and 6) some pollutants, such as the oxides of sulfur, have received more attention than other pollutants, such as the oxides of nitrogen, and thus important effects may be concealed from us by our own inattention.

To repeat, toxicologists have clearly shown us that the urban atmosphere contains a great many toxic materials with considerable potential for harm, and epidemiologists have clearly shown us that the environment contributes to disease. Almost without exception, whatever index of adverse effect on health is employed, a harmful effect of the polluted environment can be demonstrated for man.

But, and very importantly, the more specifically one attempts to pin the adverse effect on a single constituent of the atmosphere, the weaker the evidence becomes. To put it another way, air pollution clearly and consistently can be shown to have an effect on health, but the effect on populations of specific air pollutants in the urban atmosphere consistently escapes us.

In the past we have excused our difficulties by saying that the problem is multifactorial. The essence of the concept of multifactorial causes is that each environmental agent is considered to cause an effect or class of effects that is specific to the agent, but modified by other factors within the environment or host. That is, if by some trick, one could hold everything stable and had the analytic tools, one would still be able to identify the agent and its specific effect, albeit modified. For example: when sulfur dioxide enters the atmosphere, although it may react with other pollutants, there is still the specific agent, SO_2 or its derivatives, present to act against the host. Further, it acts in a specific manner within the host although the response is modified by factors within the host. Multifactoriality implies that cause and effect are specific but that there is interference or enhancement by other factors.

I would like to suggest that air pollution, and certain other phenomena that simultaneously affect health, such as poverty, constitute not simply a set of multifactorial variables, but rather a distinctly different grouping. We have chosen to call this different grouping that of multiplex variables. A multiplex variable, although made up of conceptually discrete parts, acts as a whole in terms of its effects. Multiplex variables have characteristics which make them unsuitable for consideration in the way that we generally analyze or deal with cause and effect situations.

The multiplex variable is a real entity with several distinguishing characteristics: 1) The effect of the whole is greater than the effect of the sum of the known parts; 2) The relationship of the parts to each other in a multiplex variable is not fixed; 3) The manner in which the relationships of the parts vary is not fixed; 4) Time is one of the parts.

In a multiplex variable the use of a part or parts to stand for the whole (e.g., SO_2 to stand for air pollution), in attempting either to understand (analyze) or control (legislate) the whole, will not (not, may not) be successful because it is not the part that produces the effect, but rather the whole. In a previous discussion we have shown why classical cause and effect reasoning fails in the analysis of the multiplex variable. Essentially it is not only because analytic methods tend to produce distortion of effects, or undue recognition of one or another of the parts, but because the variable does not act as a set of parts, but rather as a whole.

Even when we understand that multiplex variables require a change in outlook, from even the sophisticated stage of recognizing multiple causality, we have not solved the problem. The concept still has to be applied to the practical question of how and what the critical areas are that must be controlled if an optimum environment is to be attained and adverse effects on health minimized.

In earlier days, the great successes of epidemiology involved some distinct disease or phenomenon to which some overriding causative factor could be related. For example, the relatively clear cut phenomenon of cholera could be attributed to the water supply, and yellow fever and malaria could be related to the mosquito. Control could be directed at the water, in the first example, or the mosquito in the second. The chain of causality would be completed with the discovery of the cholera vibrio, the malarial parasite, or the virus of yellow fever.

Successful epidemiology and successful control have generally followed the pattern laid out by those early achievements. It has used the simpler roads provided by a mechanistic model of cause and effect built on those achievements, and moved quickly to find and control that single factor within the environment responsible for the health effects.

The concept of the multiplex variable raises a whole host of unsolved problems for the epidemiologist because it poses analytic questions for which he has, as yet, no tools. While it may complicate life for the epidemiologist (if indeed, he accepts the concept), the concept of multiplex variables oddly enough simplifies the problem for control legislation. The concept implies that the control of single pollutants will not solve the problem -- and more importantly, may not diminish the adverse effects of air pollution on health. In concrete terms, the control of sulfur dioxide, to the degree possible, may not control the health effects of air pollution, even if SO_2 plays a part in those effects. This seeming paradox arises because we would be thinking in terms of a hypothetical mixture of specific toxicants with specific effects upon the host; whereas the effect of air pollution on health comes about from air pollution as a multiplex whole, not simply because SO_2 is a part of air pollution.

Why is this? At least in part because the host (man) has a limited number of ways in which to respond to insults. Sulfur dioxide may cause irritation of the mucous membranes (with resultant increased airway resistance, cough, diminished bacterial clearance -- or what have you), but so may the oxides of nitrogen. Mucous membrane irritation may also result from certain airborne particles; or mixtures of various gases; or gases and particles; or cold temperatures, gases, and particles. The health effect, mucous membrane irritation, is an example of a finite number of host responses, but its causes are an almost infinite number of possible combinations of environmental influences.

It is important to note that host responses seem to have thresholds -- that there is a point below which they do not occur, and another point above which it takes increasingly greater insults to make the host response even a little worse. This kind of general host response is different from the specific host responses we associate with infectious diseases (cholera, diptheria, yellow fever, etc.), or specific toxicants (beryllium, carbon monoxide, chlorine, etc.). The multiplex variable, the cause, has almost infinite possibilities, but the response of the host has finite limits.

To return to the example I have taken, reducing sulfur dioxide may actually aggravate the effects of the environment. In the real, complicated world of combustion processes and numerous sources of pollution, it is rarely possible to remove one pollutant without affecting the others. For example, a fuel with lower sulfur content may burn at a higher temperature and actually increase the amount of oxides of nitrogen; possibly not reducing the atmospheric potential for human effects.

Thus, we can see that the concept of a multiplex variable, a cause in itself although made up of conceptually discrete parts, acting as a whole in terms of its effects, denies us the possibility of controlling the environment by controlling its parts individually.

It does, however, offer a conceptual base for a very much simpler approach to the problem of control. It provides a basis in theory for the very old public health belief that a clean environment is a healthy one, and a cleaner environment a healthier one. There is, in other words, very good reason to reduce environmental pollution as much as possible without awaiting specific evidence of harm for each of the myriad of pollutants (even if that were a possibility).

There is a control concept that appears to meet the tests of reality in the way pollutants are produced, and the way they produce their effects, and that is the control of emissions to the greatest extent feasible, employing the maximum technological capability. While seemingly simplistic, the concept fits the growing body of knowledge that points to air pollution as a multiplex variable acting as a whole rather than as the sum of its known parts. In addition, determining feasibility and technological capability, although difficult, are vastly easier than unraveling the effects of the various pollutants or determining at what level an individual pollutant may be safe or unsafe. Indeed the concepts advanced earlier make it clear that such safe or unsafe levels could probably be determined with accuracy only for that non-existent entity -- a stable and unchanging atmosphere to which stable and unchanging peop'e are exposed.

When this control philosophy was first advanced several years ago, it was suggested that because of its broadness it could never receive the public and legislative support necessary for its realization -- as opposed to the enactment of individual standards for individual pollutants (to which it is antithetical). But it has become obvious from the strength of the current environmental movement that such support is present and unfulfilled by present control concepts. Moreover, current control concepts do not meet the important requirements set earlier -- they have no predictive ability.

In essence, what is suggested is process control rather than the control of individual pollutants. Process control recognizes the fact that any avoidable soiling of the environment from a process that produces emissions is undesirable. What is avoidable or not avoidable, in practical terms, is determined by feasibility and technological capability.

Further, the concept implies that no one pollutant can be controlled without thought being given to the effect of the reduction of that pollutant on the production of others. Strategies of pollutant reduction become important. It is the overall atmosphere that is the critical item. When one pollutant receives inordinate attention because of its ease of control or popular recognition, the effect of controlling that pollutant on the total atmospheric burden of other agents must also receive attention. As in the example of SO_2 cited earlier, the net effect of a control method could conceivably be an increase in health effects (to say nothing of the effects on fuel patterns, electric power availability, economics and so forth). Thus it is possible to plan pollution reduction without meaning simply individual pollutant reduction.

The preceding chapters have amply documented what a complicated problem faces us. Each more searching look seems further to mire us in the multiplicity of factors in both man and his environment that contribute to the effects on health. As an epidemiologist, it seems to me that only a re-examination of our most basic concepts of cause and effect will help extricate us from the maze. We face the same maze in the other great health problems of our time such as poverty. The concept of multiplex variables arose in response to the need for such re-examination of causality. But it does not matter whether one accepts the concept even in its present primitive stage. It is important, however, to realize that the environment cannot be dealt with piece by piece, but must be faced as the whole that it is. The time is past when we can accept the solution of yesterday's, or even today's, environmental problem as the basis for the solution to tomorrow.

Finally, I truly believe, in terms of action rather than theory, the final control of environmental pollution rests on the drawing boards of the young who see all pollution as bad simply because it is pollution. For them the design of no technological process is complete or acceptable as long as it remains a polluting process. History tells us that was the way we were freed of the great infectious scourges of the past and that is the way we will return to harmony with our environment.

CHAPTER 17. CONCLUSIONS AND RESERVATIONS

Douglas H. K. Lee,
National Institute of Environmental Health Sciences
Research Triangle Park, North Carolina

RECAPITULATION

Since this book is intended primarily for the non-specialist, it is perhaps permissible for a generalist to review and summarize the individual contributions, each by an authority in his field.

The pervasive conditions of change -- continuing change; change in both the population at risk and the risks themselves -- is emphasized in the introductory chapter. Happily so, because each subsequent contributor deals more with the current cross-section of events within his purview than with the disturbing hand of time. The present public interest in environmental affairs is a third element of change, which one hopes will continue and accelerate improvement. Change is indeed a problem for all. A fourth element, evident to anyone a few years away from graduation, is the tremendous recent increase in detailed knowledge of biological structure and function that must be taken into account when one attacks biomedical problems such as respiratory disease. Chapter 2 presents such details for the lung to which others repeatedly refer in making their particular points.

Chapters 3 and 4 review the responses of the two major subdivisions of the respiratory tract to incident stimuli and the defense mechanisms at their disposal for meeting insults. The point is made, many times confirmed in subsequent contributions, that the number of responses available is relatively small, whereas the potential threats are legion. The alveoli, though screened from many environmental agents by the antecedent air passages, are extremely fragile, precariously balanced, and vulnerable. Omniverous macrophages provide a resolute defense, but their capabilities are limited; fibroblastic strengthening of the walls may keep the invader out, but in doing so upset the balance of function. The thick-walled air passages, adsorbing and trapping invading particles and even gases, are themselves exposed to relatively high concentrations, from which they may suffer severely in consequence. The terminal and respiratory bronchioles, situated between the two and enjoying some of the advantages and disadvantages of both, have their own problems as well that render them particularly liable to attack.

Chapter 5 discusses in eminently clear fashion four main types of disturbance that may arise from the reactions just reviewed: restriction of inspiration through increased stiffness of the lungs; restriction of expiration through airway obstruction; maldistribution of air and blood in the lungs; and impairment of oxygen diffusion. All may contribute to the symptom of shortness of breath, but it is failure of gas exchange that kills.

The range of airborne contaminants that may affect the lung are reviewed in Chapter 6 under the categories of urban, occupational, personal, and natural pollution. The variability of contaminant patterns and of the exposed lungs' reactivity become eminently apparent. The wonder is that any reasonable classification of etiology or of effect is possible at all. Adding to the variability, Chapter 7 reminds us that environmental agents may also arrive indirectly at the lung, having gained entrance to the body through the alimentary canal, the skin, or even through the chest wall. The lung, moreover, is to some extent at the mercy of countless metabolic and disease products formed in other tissues of the body.

In view of the enormous opportunity for variable responses to environmental situations, the task assigned to the author of Chapter 8, that of describing specific effects, was most formidable. That he was able to indicate some fairly clear instances approaching a one-to-one relationship is a tribute to his perspicacity; that in so doing he served mainly to underscore the usual multiplicity and variability of response was inevitable.

The pathologist, pursuing step by step in Chapter 9 the disturbances brought about by air pollutants, progressively emphasizes the virtual impossibility of establishing any great array of specific effects. Broad categories of effect, such as chronic bronchitis, emphysema, intrinsic alveolitis, and fibrosis can be established, but individual pollutants enter in very mixed fashion into their causation, and the various categories may develop side by side in the one patient in a wide variety of patterns.

In Chapter 10 we return once more to the variability of the target organ, this time in terms of its responses to environmental agents, and find our convictions amply confirmed. Hitherto we have been deliberately avoiding one large class of environmental agents -- the micro-organisms -- while examining the basis of reaction to chemical agents. Introduced at this point, they represent yet another item in an already complex situation. In general, air pollutants would seem to decrease the resistance of pulmonary tissues to infection (Chapter 11), as might be expected. It is interesting to note, however, that the balance can go the other way if the dosage and timing are just right. The reminder that associations do not always work in additive fashion is salutary, consequences are sometimes less than feared, albeit fortuitously. Two other useful caveats appear in this chapter: that the "tolerance" effects of prior exposure to oxidant pollutants may be limited; and that the infectivity of an organism for the species used is critical in interpretation of experimental studies on infectious agent interactions.

The ubiquitous factor of smoking is taken up in Chapter 12, specifically as it enters into the causation of lung cancer. That this personal pollution shows high ascendancy over community air pollution as a carcinogenic agent does not mean of course, that the latter is entirely blameless; nor, of course, is air pollution's importance in the development of other pulmonary diseases to be minimized. Chapter 13 turns the spotlight on to occupational exposures as providing information on pulmonary reactions to specific agents, and early information of situations to which the general community may be exposed, although in smaller doses, when industrial products are subsequently distributed. Here, also, the confounding and often synergistic effect of cigarette smoking in the study of pulmonary responses to occupational exposures is well brought out.

In discussing the relationship of asthma to air pollution, Chapters 14 and 15 bring out the point that epidemics of asthma occurring in a restricted locality are attributable to the dissemination of a specific (although not always identified) agent under suitable meteorological conditions. With non-epidemic asthma, on the other hand, symptoms may be exacerbated by air pollutants but the disease is attributable to allergens of more or less natural origin.

Chapter 16 takes an iconoclastic although refreshing line in decrying not only the unitary approach, but also the developing multifactorial approach to control of environmental agents. The author makes the point that a given complex of environmental agents has an individuality and impact of its own over and above those of the several agents of which it is comprised. The practical application of this philosophy -- clean up the environment to the maximum feasible extent, regardless of components -- may be more difficult to execute than to advise, and the repercussions of such sweeping change would also need examination; even with specific agents, proposed cures have sometimes turned out to be as suspect as the cause.

DISTILLATION

For the author of this final chapter, the messages that come through clearly from what has gone before may be summarized as follows.

    1. A wide range of environmental agents, even apart from infections, can participate in the causation, and even more markedly in the

aggravation, of respiratory disease. Man's ingenuity in the creation of new substances inadvertently increases the range of agents that may affect him. His dependence upon natural "sinks" to absorb or dilute these additions is proving fallacious.

2. Environmental agents seldom act singly. Stack effluents contain many potentially toxic gases as well as particulates; automobile exhausts add a variety of oxidants and polycyclic hydrocarbons; domestic activities bring their quota of suspect materials; and smoking caps the list for a large segment of the population.

3. Acute responses to relatively high doses of a toxic agent are often dramatic or even lethal. But the lesser effects of lower doses, particularly when repeated over long periods of time, may in the long run be of greater significance for the productivity and welfare of an exposed population, or even of the individual. Acting imperceptibly, producing only small immediate changes in function, or eliminating only small segments of the lung at any one time, they do not draw attention to their operation until a considerable amount of damage has occurred.

4. To meet a very large and growing range of potential insults, the respiratory system has a relatively limited number of defensive mechanisms available. It is only occasionally, therefore, that a well-defined, specific chronic disease can be linked to one specific agent. For the most part, the various agents contribute, each in its own fashion, to such generalized disturbances as chronic bronchitis, emphysema, fibrosis, intrinsic alveolitis, and cancer. It is well nigh impossible to argue backwards from a recognized chronic disease state to a specific cause; about the best that can be hoped for is epidemiological association of disease prevalence with exposure to some category of

pollution. Even such apparent exceptions as berylliosis include many cases where the diagnosis is a matter of probability rather than certainty.

5.   The pulmonary tissues that become the targets for environmental agents, and the host in general, are in a constantly dynamic state, subject to widely fluctuating physiological states, as well as a variety of non-environmental influences. This variability changes both the susceptibility and reaction to environmental agents, adding to the variability of response and the difficulty of associating individual agents with specific reactions.

6.   The long time lag between commencement of exposure and the appearance of frank disease mentioned above, increases the difficulty of pinpointing the particular agents responsible for a given disease state in an individual. Determination of exact causes is important for deciding what should be controlled, and often enters into legal questions of compensation, but the trail is often cold and incrimination a matter of presumption.

7.   The very nature of the lung's chronic responses -- fibrosis, destruction of tissue -- makes it unlikely that much recovery can be expected, even when exposure is stopped. Some functional recovery can occur as the patient learns to use his remaining capacities more effectively and eschews avoidable additional insults (smoking, infection). Repair processes can restore damaged genetic material if given a chance and thus reduce the probability of cancer. But, in general, no more than a modest improvement in health can be expected.

IMPLEMENTATION

Environmental scientists are sometimes faulted for not applying their findings to the practice of environmental control. In general, of course, the person trained in research is not well equipped to deal with the engineering, economic, or legal aspects of control activities. Nevertheless, he can be expected to indicate the lines of action that appear warranted, to comment on their relative urgency, and to insist that possible biomedical repercussions be taken into account in the decision process. Implementation, like change (Chapter 1), is a problem for all. In this vein the following comments are offered on the application of current knowledge to reduction of environmentally produced lung disease.

1. Three aspects of the clinician's traditional responsibility assume great importance in his handling of chronic respiratory disease caused or aggravated by environmental factors. The first is early recognition of a reduction in respiratory function, for which an increasing array of tests are becoming available (Chapter 5). The second is, as always, alleviation of symptoms that are the patient's primary concern. The third, conservation of remaining function, assumes greater importance as cure, in the traditional sense of the word, becomes remote. This last aspect may involve some readjustment in doctor-patient relations as the chances for quick clinical success recede. To these responsibilities, of course, must be added that of an enlightened citizen in promoting measures designed to reduce the incidence of preventible disease.

2. The improbability of cure strengthens the importance of prevention by restricting emission of pollutants (from stack gases to cigarette smoke), locating emissions away from population centers, or protecting the individual (through protective enclosures or respiratory filters). The strategy will vary from the multiplex approach advocated in Chapter 16 to removal of specific substances where these can be incriminated as prime offenders. Substitution of a safe product, procedure, or emission is an attractive solution if one can be sure that the substitute does not introduce problems of its own. We have witnessed instances where the deed was not as good as the

intent. It is probably in this regard that the scientist can be of greatest value in effective implementation.

3. While the blanket, multiplex, clean-it-up, direct approach has merits where the need for action is urgent, it can be unnecessarily costly, and perhaps not overly efficient, where there is time for a more sophisticated and more sharply targeted attack. For such to be possible, however, precise knowledge is essential -- knowledge of what the principal hazard is, of how it brings about undesired effects, of ways in which it can be controlled, and of socio-economic repercussions of proposed control methods. This more deliberate approach is equally applicable to the complex as to the single agent. In many respects we still lack the knowledge necessary for such critically directed action. Many of our control measures are still aimed at single substances without really adequate assurance that they are the prime offenders. The unpleasant is not necessarily dangerous; the unseen is not necessarily safe.

4. Current knowledge of what agents actually affect respiratory function, and of how they act, is still fragmentary. It may be likened to islands of hard stand surrounded by soft and swampy terrain on which it is hazardous to tread. Structures of any size reared on such uncertain foundations are not likely to be very durable. By an appearance of solidity they may entice the unwary to disastrous overload. Until the islands are extended, or some reasonable support provided in the areas of uncertainty, it behoves us to proceed cautiously, even experimentally. Inaction, on the other hand, is unrealistic and unacceptable. It is the scientist who must take the lead in extending the hard stand, developing supplementary support, and limiting the load to the bearing strength of the foundation.

5. Because of the need for immediate action, or because of pressure from a newly awakened public, agencies responsible for control measures are sometimes hurried into decisions or pronouncements that go beyond the support available from existing evidence. It can occasionally happen, under the press of circumstance, that two agencies may make pronouncements that are not entirely compatible. (The author has been recently subjected to some strong campus quizzing on this point.) Some machinery for more judicious and more consistent action seems desirable. This thought has been touched upon by no less than the President's Scientific Adviser:

> " . . . One thing that is missing is a credible group which can lay out in terms understandable to the public, Congress, and the Executive Branch too, what the scientific and technological facts are and to do it in an unbiased and credible way.
>
> The National Academy's function comes close to this, but it may not be completely adequate simply because it operates under Federal charter. I believe we need groups which can speak in an unbiased straightforward way without the kind of adversary relationships which you inevitably run into with these complex questions.
>
> We do not now have a group other than the National Academy which tries to perform this function. It may be that one is needed. Or it may be that the National Academy's functions in this regard need to be expanded. I do believe the scientific and engineering community itself needs to speak in more reasoned and rational ways about national problems. We need to develop a means of generating unbiased, authoritative positions on subjects which involve science and technology.

There are various mechanisms which you can think of to do that. Presidential commissions are one such possibility. The Academy is one such possibility. The output from the President's Science Advisory Committee is another, but all of these have shades of advocacy about them. The best you can do is get a balance of interests rather than have no conflicts of interests at all in such groups."

Science 174:(4014), 1109

To this the author would only add a suggestion that is less far out and more seriously given than may appear at first sight -- why not create an environmental judiciary, akin to the Supreme Court, financially and politically independent of possible pressures, to weigh the evidence and pass judgment on matters that might otherwise become the football of opposing interests?